Reviews of Environmental Contamination and Toxicology

VOLUME 194

Reviews of Environmental Contamination and Toxicology

Editor
David M. Whitacre

Founding Editor
Francis A. Gunther

VOLUME 194

Springer

DR. DAVID M. WHITACRE, *Editor*
Reviews of Environmental Contamination and Toxicology

5115 Bunch Road
Summerfield North, Carolina 27358, USA
(336) 634-2131 (PHONE and FAX)
E-mail: dmwhitacre@triad.rr.com

DR. HERBERT N. NIGG, *Editor*
Bulletin of Environmental Contamination and Toxicology

University of Florida
700 Experiment Station Road
Lake Alfred, Florida 33850, USA
(863) 956-1151; FAX (941) 956-4631
E-mail: hnn@LAL.UFL.edu

DR. DANIEL R. DOERGE, *Editor*
Archives of Environmental Contamination and Toxicology

7719 12th Street
Paron, Arkansas 72122, USA
(501) 821-1147; FAX (501) 821-1146
E-mail: AECT_editor@earthlink.net

Springer
New York: 233 Spring Street, New York, NY 10013, USA
Heidelberg: Postfach 10 52 80, 69042 Heidelberg, Germany

ISBN: 978-1-4419-2573-2 e-ISBN: 978-0-387-74816-0
ISSN 0179-5953

Printed on acid-free paper.

springer.com

Foreword

International concern in scientific, industrial, and governmental communities over traces of xenobiotics in foods and in both abiotic and biotic environments has justified the present triumvirate of specialized publications in this field: comprehensive reviews, rapidly published research papers and progress reports, and archival documentations. These three international publications are integrated and scheduled to provide the coherency essential for nonduplicative and current progress in a field as dynamic and complex as environmental contamination and toxicology. This series is reserved exclusively for the diversified literature on "toxic" chemicals in our food, our feeds, our homes, recreational and working surroundings, our domestic animals, our wildlife and ourselves. Tremendous efforts worldwide have been mobilized to evaluate the nature, presence, magnitude, fate, and toxicology of the chemicals loosed upon the earth. Among the sequelae of this broad new emphasis is an undeniable need for an articulated set of authoritative publications, where one can find the latest important world literature produced by these emerging areas of science together with documentation of pertinent ancillary legislation.

Research directors and legislative or administrative advisers do not have the time to scan the escalating number of technical publications that may contain articles important to current responsibility. Rather, these individuals need the background provided by detailed reviews and the assurance that the latest information is made available to them, all with minimal literature searching. Similarly, the scientist assigned or attracted to a new problem is required to glean all literature pertinent to the task, to publish new developments or important new experimental details quickly, to inform others of findings that might alter their own efforts, and eventually to publish all his/her supporting data and conclusions for archival purposes.

In the fields of environmental contamination and toxicology, the sum of these concerns and responsibilities is decisively addressed by the uniform, encompassing, and timely publication format of the Springer triumvirate:

Reviews of Environmental Contamination and Toxicology [Vol. 1 through 97 (1962–1986) as Residue Reviews] for detailed review articles concerned with any aspects of chemical contaminants, including pesticides, in the total environment with toxicological considerations and consequences.

Bulletin of Environmental Contamination and Toxicology (Vol. 1 in 1966) for rapid publication of short reports of significant advances and discoveries in the fields of air, soil, water, and food contamination and pollution as well as methodology and other disciplines concerned with the introduction, presence, and effects of toxicants in the total environment.

Archives of Environmental Contamination and Toxicology (Vol. 1 in 1973) for important complete articles emphasizing and describing original experimental or theoretical research work pertaining to the scientific aspects of chemical contaminants in the environment.

Manuscripts for *Reviews* and the *Archives* are in identical formats and are peer reviewed by scientists in the field for adequacy and value; manuscripts for the *Bulletin* are also reviewed, but are published by photo-offset from camera-ready copy to provide the latest results with minimum delay. The individual editors of these three publications comprise the joint Coordinating Board of Editors with referral within the Board of manuscripts submitted to one publication but deemed by major emphasis or length more suitable for one of the others.

Coordinating Board of Editors

Preface

The role of *Reviews* is to publish detailed scientific review articles on all aspects of environmental contamination and associated toxicological consequences. Such articles facilitate the often-complex task of accessing and interpreting cogent scientific data within the confines of one or more closely related research fields.

In the nearly 50 years since *Reviews of Environmental Contamination and Toxicology* (formerly *Residue Reviews*) was first published, the number, scope and complexity of environmental pollution incidents have grown unabated. During this entire period, the emphasis has been on publishing articles that address the presence and toxicity of environmental contaminants. New research is published each year on a myriad of environmental pollution issues facing peoples worldwide. This fact, and the routine discovery and reporting of new environmental contamination cases, creates an increasingly important function for *Reviews*.

The staggering volume of scientific literature demands remedy by which data can be synthesized and made available to readers in an abridged form. *Reviews* addresses this need and provides detailed reviews worldwide to key scientists and science or policy administrators, whether employed by government, universities or the private sector.

There is a panoply of environmental issues and concerns on which many scientists have focused their research in past years. The scope of this list is quite broad, encompassing environmental events globally that affect marine and terrestrial ecosystems; biotic and abiotic environments; impacts on plants, humans and wildlife; and pollutants, both chemical and radioactive; as well as the ravages of environmental disease in virtually all environmental media (soil, water, air). New or enhanced safety and environmental concerns have emerged in the last decade to be added to incidents covered by the media, studied by scientists, and addressed by governmental and private institutions. Among these are events so striking that they are creating a paradigm shift. Two in particular are at the center of ever-increasing media as well as scientific attention: bioterrorism and global warming. Unfortunately, these very worrisome issues are now super-imposed on the already extensive list of ongoing environmental challenges.

The ultimate role of publishing scientific research is to enhance understanding of the environment in ways that allow the public to be better informed. The term "informed public" as used by Thomas Jefferson in the age of enlightenment conveyed the thought of soundness and good judgment. In the modern sense, being "well informed" has the narrower meaning of having access to sufficient information. Because the public still gets most of its information on science and technology from TV news and reports, the role for scientists as interpreters and brokers of scientific information to the public will grow rather than diminish.

Environmentalism is the newest global political force, resulting in the emergence of multi-national consortia to control pollution and the evolution of the environmental ethic. Will the new politics of the 21st century involve a consortium of technologists and environmentalists, or a progressive confrontation? These matters are of genuine concern to governmental agencies and legislative bodies around the world.

For those who make the decisions about how our planet is managed, there is an ongoing need for continual surveillance and intelligent controls, to avoid endangering the environment, public health, and wildlife. Ensuring safety-in-use of the many chemicals involved in our highly industrialized culture is a dynamic challenge, for the old, established materials are continually being displaced by newly developed molecules more acceptable to federal and state regulatory agencies, public health officials, and environmentalists.

Reviews publishes synoptic articles designed to treat the presence, fate, and, if possible, the safety of xenobiotics in any segment of the environment. These reviews can either be general or specific, but properly lie in the domains of analytical chemistry and its methodology, biochemistry, human and animal medicine, legislation, pharmacology, physiology, toxicology and regulation. Certain affairs in food technology concerned specifically with pesticide and other food-additive problems may also be appropriate.

Because manuscripts are published in the order in which they are received in final form, it may seem that some important aspects have been neglected at times. However, these apparent omissions are recognized, and pertinent manuscripts are likely in preparation or planned. The field is so very large and the interests in it are so varied that the Editor and the Editorial Board earnestly solicit authors and suggestions of underrepresented topics to make this international book series yet more useful and worthwhile.

Justification for the preparation of any review for this book series is that it deals with some aspect of the many real problems arising from the presence of foreign chemicals in our surroundings. Thus, manuscripts may encompass case studies from any country. Food additives, including pesticides, or their metabolites that may persist into human food and animal feeds are within this scope. Additionally, chemical contamination in any manner of air, water, soil, or plant or animal life is within these objectives and their purview.

Manuscripts are often contributed by invitation. However, nominations for new topics or topics in areas that are rapidly advancing are welcome. Preliminary communication with the Editor is recommended before volunteered review manuscripts are submitted.

Summerfield, North Carolina D.M.W.

Contents

Microbially Derived Off-Flavor from Geosmin and 2-Methylisoborneol: Sources and Remediation

Kishore Kumar Krishnani[1], Pitchaiyappan Ravichandran[1], and Subbanna Ayyappan[2]

I Introduction

The occurrence of tastes and odors is a recurrent problem in the beverage, potable water, food, and aquaculture industries. Taste-and-odor (T/O) occurrences have been documented in a number of public water supply reservoirs (Silvey et al. 1950; Suffet et al. 1996) in the United States (Rosen et al. 1970; Izaguirre et al. 1982; Seligman et al. 1992; Burlingame et al. 1986, 1992; Young et al. 1999), Canada (Slater and Blok 1983a, b), Japan (Yagi et al. 1981, 1983; Yagi 1988; Hosaka et al. 1995), Australia (Hayes and Burch 1989; Baker 1992; Baker et al. 1994, 2001), The Netherlands (Piet et al. 1972), Sweden (Lundgren et al. 1988), Germany (Mohren and Jüttner 1983), Finland (Veijanen et al. 1988, 1992), France (Cotsaris et al. 1995), India (Desikachary 1959; Arora and Gupta 1983; Krishnani et al. 2005, 2006a),

[1] Central Institute of Brackishwater Aquaculture, 75, Santhome high road, R.A. Puram, Chennai-600028, India

[2] Krishi Anusandhan Bhavan, Indian Council of Agricultural Research, Pusa, New Delhi, 110012, India

Communicated by G.W. Ware.

Taiwan (Lin et al. 2002), and Spain (Sabater et al. 2003; Vilalta et al. 2003, 2004). Food industries, including grapes (La Guerche et al. 2005), apples (Frank 1977; Siegmund and Pollinger-Zierler 2006), pears (Nunes 2002), peaches (Mercier and Jimenez 2004), and vegetables such as dried beans or beetroot (Maga 1987) have also been affected with inconsistent flavor. Based on suspected origins, Tucker and van der Ploeg (1999) and van der Ploeg (1991) categorized off-flavors as rotten, decayed, cardboard, stale, petroleum, fishy, woody, earthy or muddy, and musty. This chapter presents an extensive review of chemical causes of off-flavor problems, especially with reference to muddy and musty flavor in aquatic organisms and possible remediation techniques.

II Chemical Causes of Off-Flavor

Problems of off-flavors caused by chemicals have been reported for fish (Whitfield et al. 1994) and Crustacea (Whitfield et al. 1981, 1988). Petroleum off-flavors occur mainly from accidental spills of diesel fuel when fish are exposed to persistent petroleum products, causing long-lasting flavor problems (Tucker and van der Ploeg 1999; Motohiro 1983). Rotten and sulfury off-flavors have been attributed to polysulfides formed by decomposition of blooms in freshwater reservoirs in Australia (Hayes and Burch 1989). Dimethyl trisulfide has been correlated with off-flavor problems in cheese, prawns, and vegetables (Hayes and Burch 1989; Whitfield et al. 1981).

Algal volatile organic compounds (AVOCs), mainly terpenoids, cause economic losses to water, food, and aquaculture industries because of reported taste and odor (Wnorowski and Scott 1992; Engle et al. 1995; McGuire 1995; Watson 1999, 2003), which establish chemical communication among organisms (Harborne and Tmoas-Barberan 1989; Harrawijn et al. 2001). The role of natural biofilm inside pipelines as a potential source and reservoir for odorous volatile organic compounds has been well documented (Skjevrak et al. 2005). Watson and Ridal (2003) found that periphyton is a major T/O source in the St. Lawrence River.

A large group of compounds such as 2-methoxy-3-isopropylpyrazine, di-, tri-, tetra-, and pentachloro anisoles, octa-1,3-diene, 2-methylisoborneol, (MIB), and geosmin are responsible for an earthy-musty off-flavor (Kilkast 1993). Schnurer et al. (1999) characterized fungal volatiles from mainly *Aspergillus*, *Fusarium*, and *Penicillium* with gas chromatography, mass spectrometry, and sensory analysis; common volatiles found were 2-methyl-1-propanol, 3-methyl-1-butanol, 1-octen-3-ol, 3-octanone, 3-methylfuran, ethyl acetate, and the malodorous 2-methylisoborneol and geosmin. Fravel et al. (2002) characterized volatile compounds emitted by sclerotia of *Sclerotinia minor* and *Sclerotinia sclerotiorum* and identified as 2-methylenebornane and 2-methylisoborneol by solid-phase microextraction followed by gas chromatography and mass spectrometry.

Because this review focuses mainly on geosmin (*trans*-1,10-dimethyl-*trans*-9-decalol) and 2-MIB (1,2,7,7 tetramethyl-*exo*-bicyclo heptan-2-ol), their chemical

Table 1 Physical and chemical characteristics of geosmin and 2-methylisoborneol (MIB)

Characteristics	Geosmin	2-Methyl isoborneol
Structure	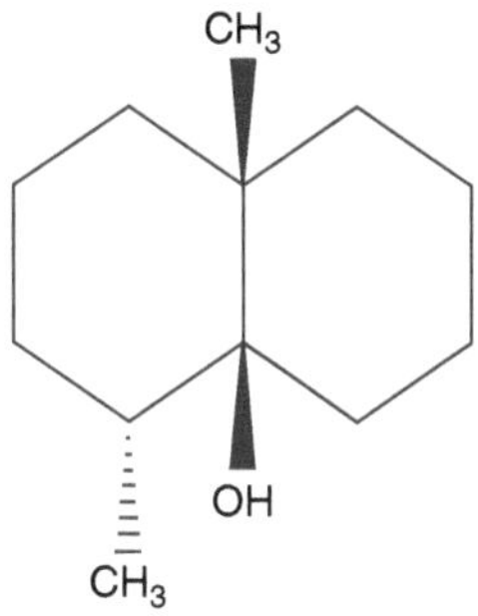	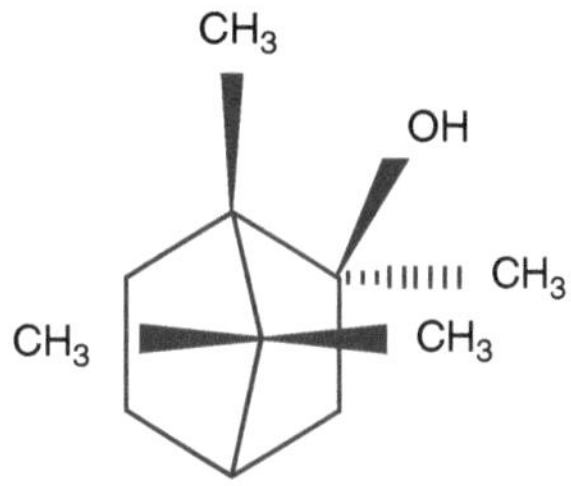
Chemical name	*trans*-1,10-Dimethyl-*trans*-9-decalol	1,2,7,7-Tetramethyl-exo-bicyclo-heptan-2-ol
Molecular formula	$C_{12}H_{22}O$	$C_{11}H_{20}O$
Formula weight	182.31	168.28
Appearance	Light yellow oil	White solid
Boiling point	270°C	—
Organoleptic properties	Earthy-muddy	Musty
Odor threshold concentration (ng/L)	10	29
Human olfactory sense (ng/L)	4	7–15
Toxicity to rainbow trout (mg/L)	0.45	10
LC_{50} (sea urchin embryos) (mg/L)	17	69

Sources: Cees et al. 1974; Gagne et al. 1999; Gerber 1968, 1969; Nakajima et al. 1996; Persson 1979; Watson et al. 2000.

structures and chemical and physical characteristics are shown in Table 1. Geosmin and MIB implicated in muddy-musty flavors of water and fish are a problem in the aquaculture, food, beverage, and potable water industries. These compounds are reported as a source of musty-earthy flavor in grain caused by improper storage (Wasowicz et al. 1988; Jelen et al. 2003). Geosmin is also known to contribute to a characteristic earthy red beet flavor (Lu et al. 2003) and an earthy smell in grapes (La Guerche et al. 2005). MIB was related to musty-earthy notes in Brie and Camembert cheese flavor (Karahadian et al. 1985). Siegmund and Pollinger-Zierler (2006) detected higher concentrations of 2-isopropyl-3-methoxypyrazine, 2-isobutyl-3-methoxypyrazine, 2-methylisoborneol, 1-octen-3-ol, fenchyl alcohol, geosmin, and guaiacol as well as 2,6-dibromophenol in apple juice samples.

Geosmin and MIB have been reported to be responsible for most source-water odors (Persson 1979, 1981, 1983, 1988; Wnorowski 1992; Wnorowski and Scott 1992; Saxby 1993; Jüttner 1995). The FLAVOR Profile (FPA) method was adapted to drinking water by Krasner et al. (1985). The FPA method was also employed by Lin et al. (2002) to determine the odor groups in the source water of two water treatment plants in Taiwan, and chemical analysis showed that MIB and geosmin were present in the source water and were responsible for the musty odor.

MIB and geosmin are stereoisomeric and can be detected easily at low levels by human olfactory senses because of their very low odor threshold concentration

(Cees et al. 1974; Tyler et al. 1978; Persson 1979; Polak and Provasi 1992; Veijanen et al. 1988, 1992; Watson et al. 2000). A Lab test conducted by Jung et al. (2004) revealed that threshold odor levels for MIB and geosmin appeared to be 30 ng/L, which can create consumer complaints. However, concentrations as low as 10 ng/L can impart off-flavors to a variety of food and water sources (Krasner et al. 1985; Dionigi et al. 1993). Concentrations as low as 10 ng/L in water and 0.7 μg/kg in fish can be detected (Zimmerman et al. 2002).

Occurrence of geosmin and MIB are common in aquaculture as bioaccumulation of these sesquiterpenoids in fish and shellfish causes off-flavors in farmed and wild stocks (Persson 1981; Lovell and Broce 1985; Hsieh et al. 1988; Dionigi et al. 1998; Farmer et al. 1995; Lazur 2004). During summer months with higher feeding rates, conditions are conducive to the growth of certain species of algae and bacteria, causing off-flavor in fish and shellfish and making them unmarketable (Kajino and Sakamoto 1995; Eynard et al. 2000). Tellez et al. (2001a) concluded that besides MIB several volatile compounds may cause off-flavor problems in catfish aquaculture; however, off-flavor due to MIB may mask the odors of dimethyl disulfide/trisulfide. Nakajima et al. (1996) and Gagne et al. (1999) reported toxicities levels of geosmin and MIB for sea urchin embryos (*Hemicentrotus pulcherrimus* Agassiz) and rainbow trout (*Oncorhynchus mykiss* Walbaum) (see Table 1).

III Geosmin- and MIB-Producing Species

Geosmin, an earthy-smelling substance, was isolated in 1964 by Gerber and Lechevalier (1965). MIB, a musty- or camphorous-smelling compound, was reported in 1969 by Medsker et al. (1969) and Gerber (1969), and independently by Rosen et al. (1970) in 1970. Geosmin and MIB were first identified in actinomycetes (Gerber 1968, 1969, 1979, 1983; Blevins 1980; Yagi et al. 1981, 1983; Bentley and Meganathan 1981; Schrader and Blevins 1993), then later in cyanobacteria (Izaguirre et al. 1982; Wu and Jüttner 1988; Martin et al. 1991; Matsumuto and Tsuchiya 1988; Tsuchiya et al. 1981; Tsuchiya and Matsumoto 1988; Schrader and Blevins 1993; Tabachek and Yurkowski 1976) and fungi (Kikuchi et al. 1981) that inhabit aquatic and soil environments (Tables 2–4).

Siegmund and Pollinger-Zierler (2006) reported for the first time the presence of *Streptomyces* sp. as the spoilage bacteria of apple juice. Initially, only certain actinomycetes were reported to produce MIB; later, various cyanobacterial species from the genera *Anabaena, Oscillatoria, Lyngbya,* and *Phormidium* have been reported to produce musty and earthy flavors in cultured catfish (Tucker 2000). MIB-producing *Lyngbya* species was reported from Manitoba fish farming (Tabachek and Yurkowski 1976; Yurkowski and Tabachek 1980). In catfish ponds, MIB is usually produced by the blue-green alga *Oscillatoria perornata* (*Planktothrix perornata*) (Tucker 2000). Martin et al. (1988) were the first to report MIB-related off-flavor in commercial farm-raised channel catfish and later attributed it to a planktonic *Oscillatoria* sp. (Martin et al. 1991). Tellez et al. (2001a) reported major

Table 2 MIB-producing species

Species	Origin	Habitat	References
Oscillatoria			
O. perornata (Planktothrix MS988)	Fish pond/USA	Planktonic	van der Ploeg et al. 1995; Tellez et al. 2001a, b; Taylor et al. 2006
O. limosa	Lake/USA	Benthic	Izaguirre and Taylor 1995
Oscillatoria sp.	Fish pond/USA	Planktonic	Martin et al. 1991
O. tenuis	Japan	Planktonic	Negoro et al. 1988
O. geminata	Fish pond/Japan	Fish Pond	Matsumoto and Tsuchiya 1988
O. limnetica	Fish pond/Japan	Fish Pond	Matsumoto and Tsuchiya 1988
Oscillatoria cf. *curviceps*	Lake/USA	Benthic	Izaguirre et al. 1982, 1983
O. tenuis	Water supply/USA	Benthic	Izaguirre et al. 1983
O. variabilis	Fish farming lake/ Japan	Benthic	Tabachek and Yurakowski 1976
O. chalybea	Reservoir/ Israel	Benthic	Leventer and Eren 1970
Phormidium			
Phormidium LP684	Lake/USA	Benthic	Taylor et al. 2006
Phormidium aff. formosum	Water supply/ Australia	Benthic	Baker et al. 2001
P. favosum	Lake/Japan	Benthic	Sugiura et al. 1997
Phormidium	USA	Benthic	Izaguirre 1992
P. tenue	Lake/Japan	Benthic	Sugiura et al. 1986
P. tenue	Water supply/ Japan	Planktonic	Yagi et al. 1983
Pseudanabaena			
Pseudanabaena	Reservoirs/USA	Planktonic	Izaguirre et al. 1999; Taylor et al. 2006
Pseudanabaena	Lake/USA	Planktonic	Izaguirre and Taylor 1998
Other species			
Synechococcus sp.	Water reservoirs/USA	Planktonic	Taylor et al. 2006
Leptolyngbya sp.	Periphyton, lake/USA		Taylor et al. 2006
Lyngbya LO198	Reservoir/USA	Benthic	Taylor et al. 2006
Hyella	Aqueduct water/USA	Epiphytic	Izaguirre and Taylor 1995
Lyngbya Cal.Aq.892	Aqueduct lake/USA	Epiphytic	Izaguirre and Taylor 1995
Planktothrix MS988	Catfish pond/ USA	Planktonic	Martin et al. 1991
Planktothrix cryptovaginata	Fish, water/Finland	Benthic	Persson 1988
Jaaginema geminatum	River/Japan	Benthic	Tsuchiya and Matsumoto 1988
Synechococcus sp.	Plankton, lake/USA	Planktonic	Izaguirre et al. 1984
Lyngbya cf. *aestuarii*	Fish farming lake/ Japan	Benthic	Yurkowski and Tabachek 1980 Tabachek and Yurkowski 1976

Table 3 Geosmin-producing species

Species	Origin	Habitat	References
Anabaena			
Anabaena sp.	Lake/USA	Planktonic	Saadoun et al. 2001
A. laxa CA 783	Lake plankton/USA	Planktonic	Rashash et al. 1996
A. crassa LS698	Lake/USA/Australia	Planktonic	Baker et al. 1994; Komarkova-Legnerova and Cronberg 1992
A. circinalis	River/Australia	Planktonic	Bowmer et al. 1992
A. circinalis	Reservoir/USA	Planktonic	Rosen et al. 1992
A. solitaria	Taiwan	Planktonic	Wu et al. 1991
A. viguieri	Taiwan	Planktonic	Wu et al. 1991
A. macrospora	River/Japan	Planktonic	Tsuchiya and Matsumoto 1988
A. scheremetievi Elenkin	Water supply/USA	Planktonic	Izaguirre et al. 1982
Oscillatoria			
O. limosa	River/Spain	Benthic	Vilalta et al. 2003, 2004
O. limosa	River/Reservoir/ Netherlands		van Breeman et al. 1992
Oscillatoria sp. (Philadelphia)	Periphyton, river/ USA	Benthic	Burlingame et al. 1986
O. brevis	Inland water/Norway	Benthic	Berglind et. al. 1983b
O. simplicissima	Water supply/USA	Pipeline	Izaguirre et al. 1982
O. tenuis	Fish pond/Israel		Aschner et al. 1967
Phormidium			
Phormidium LS1283	Algae, lake/USA	Benthic	Taylor et al. 2006
Phormidium cf. inundatum LO584	Reservoir/USA	Sediment	Taylor et al. 2006
Phormidium sp. (SDC202a,b,c)	Canal/USA		Taylor et al. 2006
Phormidium sp. DCR301	Reservoir/USA	Sediment	Taylor et al. 2006
Phormidium sp. ER0100	Reservoir/USA	Sediment	Taylor et al. 2006
Phormidium DC 699	Algae/lake/USA	Benthic	Taylor et al. 2006
Phormidium sp. LD499	Algae/ lake	Benthic	Taylor et al. 2006
Phormidium sp. LM494	Lake/USA	Sediments	Taylor et al. 2006
Phormidium sp. LS587	Lake/USA	Sediments	Taylor et al. 2006
Phormidium sp. R12	Canal/USA		Taylor et al. 2006
P. allorgei	Lake/Japan	Benthic	Sugiura et al. 1997
Phormidium sp.	Lake/USA	Benthic	Izaguirre and Taylor 1995
P. amoenum	Japan	Benthic	Tsuchiya and Matsumoto 1988
P. simplissimum	Fish, water/Finland	Benthic	Persson 1988
P. formosum	Fish, water/Finland	Benthic	Persson 1988
P. cortianum	Fish farming lake/ Japan	Benthic	Tabachek and Yurakowski 1976
Other geosmin-producing species			
Nostoc sp.	Creek/USA	Periphytic	Taylor et al. 2006
Microcoleus-like cyano	Aqueduct/USA	Epiphytic	Izaguirre and Taylor 1995

(continued)

Table 3 (continued)

Species	Origin	Habitat	References
Lyngbya cf. *subtilis*	Aquaculture pond/ USA	Benthic	Schrader and Blevins 1993
Planktothrix prolifica	Norway	Benthic	Naes et al. 1988
Aphanizomenon gracile	Lake/Germany	Planktonic	Jüttner 1984
Tychonema bornetii	Lake/Norway	Benthic	Berglind et al. 1983a
Schizothrix muellerii	Japan	Benthic	Kikuchi et al. 1973
Symploca muscorum	Fish farming lake/ Japan	Soil	Tabachek and Yurakowski 1976 (first reported by Medsker et al. 1968)
Geitlerenema splendidum	Fish farming lake/ Japan	Benthic	Tabachek and Yurakowski 1976
Actinomycetes			
Streptomyces halstedii	Aquaculture pond/ USA	Sediments	Schrader and Blevins 2001
Streptomyces griseus	USA		Gerber and Lechevalier 1965

Table 4 Geosmin- and MIB-producing species

Species	Origin	Habitat	References
Phormidium			
Phormidium sp. Cal Aq.0100	Aqueduct/USA	Periphyton	Taylor et al. 2006
Phormidium sp.HD798	Algae/lake	Periphytic	Taylor et al. 2006
Phormidium sp.	Lake/USA	Benthic	Izaguirre 1992
Phormidium sp.	River/Japan	Benthic	Matsumuto and Tsuchiya 1988
Phormidium sp.	Inland water/ Norway	Benthic	Berglind et al. 1983b
Other species			
Synechococcus sp CL792	Lake/USA	Planktonic	Taylor et al. 2006
Nostoc sp.	Water treatment plant /Taiwan		Hu and Chiang 1996
T. granulatum	Japan	Benthic	Tsuchiya and Matsumoto 1988
Planktothrix agardhii	Lake/Norway	Planktonic	Persson 1988 Berglind et al. 1983a
O. brevis			Berglind et al. 1983b
Actinomycetes			
Streptomyces	Denmark	Streams/ponds	Klausen et al. 2005
Streptomyces violaceusniger	Water supply/ Jordon	Sediment	Saadoun et al. 1997
Streptomyces sp.	USA		Gerber 1977

volatile compounds such as heptadecane (57%), MIB (29.4%), and benzaldehyde (1.2%) from unialgal continuous cultures of the cyanobacterium *Oscillatoria perornata*. In other environments, many other species of blue-green algae (Tabachek and Yurkowski 1976; Izaguirre et al. 1982; Taylor et al. 2006) and also actinomycetes (Sivonen 1982; Scholler et al. 2002) have been reported to produce MIB. Some of these species also produce toxins (Jardine et al. 1999). Most of the cyanobacteria that produce toxins are planktonic; however, microcystin-producing benthic cyanobacteria have also been reported (Izaguirre et al. 2007), which have been characterized by 16S rRNA technique. Off-flavors other than musty and earthy in catfish are woody and pine and have been attributed to the presence of unspecific cyanobacteria (van der Ploeg 1991).

Izaguirre and Taylor (2004) observed, in drinking water supplies in the U.S., that known geosmin-and MIB-producing cyanobacteria belong to genera such as *Anabaena*, *Oscillatoria*, *Phormidium*, *Lyngbya*, *Leptolyngbya*, *Microcoleus*, *Nostoc*, *Planktothrix*, *Pseudanabaena*, *Hyella*, *and Synechococcus* (see Tables 2–4). Many MIB- and geosmin-producing *Oscillatoria* strains have been isolated from water supplies in California (Izaguirre et al. 1982).

Cultures of two *Oscillatoria* strains isolated from drinking water reservoirs in California produced 60–150 μg/L MIB (Izaguirre et al. 1982). Cultures of *O. geminata* and *O. limnetica* isolated from a fish cultivation pond and a park pond in the Tokyo area produced 550 and 240 μg/L MIB, respectively (Matsumoto and Tsuchiya 1988). *Phormidium tenue*, a major cause of MIB episodes in Lake Biwa (Yagi et al. 1983) and Lake Kasumigaura, Japan (Sugiura et al. 1986, 1998), produced 120 μg/L MIB in culture (Negoro et al. 1988). *O. limnetica* is considered synonymous with *Pseudanabaena limnetica* (Anagnostidis and Komárek 1988; Baker 1992). Two MIB-producing (240–260 μg/L) cyanobacteria, *Lyngbya* strains, were isolated from a major aqueduct system in California (Izaguirre and Taylor 1995). *Lyngbya* was a comparatively strong MIB-producing species relative to other MIB producers (Martin et al. 1991; Izaguirre 1992).

Planktonic and benthic species synthesize both compounds, geosmin and MIB (see Tables 2–4). The first planktonic MIB producers were reported in Japan (Yagi et al. 1983; Negoro et al. 1988) and later in Australia (Baker et al. 1994). The first major planktonic MIB producer isolated in the U.S. was the planktonic *Oscillatoria* (*Planktothrix*) from a catfish pond in Mississippi (Martin et al. 1991). Some new strains of *Pseudanabaena* species isolated from Castic Lake, California, represented the major planktonic MIB producers isolated from drinking water in the U.S. (Izaguirre and Taylor 1998).

Izaguirre and Taylor (2004) noted that some MIB-producing cyanobacteria isolated in the U.S. have morphological analogues in other parts of the world. MIB-producing *Planktothrix* sp. (originally called *Oscillatoria*) isolated from a catfish pond in Mississippi (Martin et al. 1991; van der Ploeg et al. 1995) appears indistinguishable from an MIB-producing *Planktothrix* species in Australia (Baker 1992), and may also be related to an *Oscillatoria* cf. *raciborskii* reported in Japan (Hosaka et al. 1995), and possibly also to the *O. tenuis* reported by Negoro et al. (1988). The other example is MIB-producing *Pseudanabaena* (Izaguirre and Taylor 1998; Izaguirre et al. 1999), some strains of which are morphologically similar to the MIB producer *Phormidium tenue* in Japan, reported by Yagi et al. (1983). Izaguirre and Taylor (2004) observed that *Pseudanabaena* strains isolated in the U.S. may also be related to the MIB-producing strain of *Oscillatoria limnetica* reported by Matsumoto and Tsuchiya (1988), as this species is considered synonymous with *Pseudanabaena limnetica* by Anagnostidis and Komárek (1988) and Baker (1992).

The first reports of geosmin production by *Anabaena circinalis, Anabaena laxa*, and *Symploca muscorum* were published by Henley (1970), Rashash et al. (1996), and Medsker et al. (1968), respectively. Many species of blue-green algae and actinomycetes can produce geosmin, but in catfish ponds, the main geosmin producers are species of the blue-green alga *Anabaena*, followed by *Aphanizomenon* or *Lyngbya* (van der Ploeg et al. 1992; Schrader and Blevins 1993). A geosmin-producing *Oscillatoria* strain and one *Anabaena* species were isolated from drinking water supplies in California (Izaguirre et al. 1982). Geosmin- and MIB-producing cyanobacteria found in the U.S. also occur in other countries. In Australia, *Anabaena circinalis*, which produces geosmin along with saxitoxin, is a major problem, causing the deaths of animals (Negri et al. 1995). This species is also reported from South African reservoirs (Wnorowski and Scott 1992). *Oscillatoria splendida* (now called *Geitlerinema splendidum*) is a common geosmin producer widespread throughout the Northern Hemisphere.

Nielsen et al. (2006) found that bacterial groups within *Actinobacteria* produce the compounds geosmin and MIB, which lower the quality of surface water when used for drinking. Results indicate that combined microautoradiography and catalyzed reporter deposition (CARD-FISH) may serve as an effective tool when studying identity and activity of microorganisms within freshwater environments (Nielsen et al. 2006). Klausen et al. (2005) attributed the occurrence of the geosmin and MIB in freshwater environments to *Actinobacteria,* most of them belonging to the genus *Streptomyces* (Schrader and Blevins 1993; Zaitlin et al. 2003; Zaitlin and Watson 2006). The new species *Penicillium discolor*, frequently isolated from nuts, vegetables and cheese, also produces the moldy-smelling compounds geosmin and MIB (Frisvad et al. 1997). A correlation between the occurrence of geosmin, argosmin and heptadec-*cis*-5-ene and the presence of the cyanobacterium *Aphanizomenon gracile* was observed by Jüttner (1984). These compounds were present in spring and autumn, when *A. gracile* also occurred in the lake, but were not detected in summer, when the organism was absent.

A *Phormidium* sp. reported by Izaguirre (1992) was rare among cyanobacteria in that it could produce both MIB and geosmin (see Table 4). Five other cyanobacteria with this property have been reported: three strains in Norway (Berglind et al. 1983b), one in Japan (Matsumoto and Tsuchiya 1988), and one in Taiwan (Wu and Jüttner 1988). Schrader and Dennis (2005) reported that geosmin and MIB were implicated for earthy and musty off-flavors, respectively, in farm-raised channel catfish (*Ictalurus punctatus*) in the Southeastern U.S. MIB-producing cyanobacterium (*Oscillatoria perornata*) is present in catfish ponds in both Mississippi and Alabama Blackland Prairie (MABP), whereas geosmin was found to be more prevalent in catfish ponds in the MABP region than West Mississippi.

IV Biosynthesis of Geosmin

Bentley and Meganathan (1981) used radiogas chromatography to investigate biosynthesis of geosmin, the characteristic odoriferous constituent of *Streptomyces* species. Based on the incorporation of acetate into geosmin by strains of *S. antibioticus*, they concluded that geosmin was likely a degraded sesquiterpene. Actinomycetes, gram-positive soil bacteria *Streptomyces avermitilis* and *S. coelicolor*, produce geosmin, and germacradienol has been identified as a precursor/cometabolite of geosmin in streptomycetes and myxobacteria (Cane and Watt 2003). The *S. avermitilis* gene SAV 2163 (geoA) and *S. coelicolor* A3 (2) SCO6073 gene encodes germacradienol/geosmin synthase (Jiang et al. 2006; Gust et al. 2003). Among the sesquiterpene synthases, the 2178-bp geoA gene (SAV 2163) encodes a putative protein of 725 amino acids with a significant similarity to the *S. coelicolor* A3(2) SCO6073 2181-bp gene product encoding 726 amino acids (Gust et al. 2003; Cane and Watt 2003). Deletion of the entire SCO6073 (SC9B1.20) gene from *S. coelicolor* A3(2) results in complete loss of geosmin production (Cane and Watt 2003; Gust et al. 2003); this provides evidence that SCO6073 encodes a germacradienol synthase, which catalyzes an essential step in the biosynthesis of geosmin. *Streptomyces avermitilis* mutants with a deleted geoA were unable to produce either germacradienol or geosmin, and biosynthesis of both compounds was restored by introducing intact geoA gene in mutants (Cane et al. 2006).

Cane and Watt (2003) expressed a 2181-bp gene from *S. coelicolor* A3(2) (SCO6703 = SC9B1.20) in *Escherichia coli* to give a 726-amino-acid protein and originally proposed that formation of geosmin from germacradienol would involve multistep biochemical redox pathways catalyzed by several hypothetical enzymes, which has also been suggested by other researchers (Spiteller et al. 2002; Dickschat et al. 2005). Cane and Watt (2003) and He and Cane (2004) revealed that biosynthetic conversion of fernesyl diphosphate to geosmin requires a divalent cation, preferably Mg^{2+} and no other organic or inorganic cofactor is required. Recently, Jiang et al. (2006) successfully demonstrated that a single enzyme (germacradienol D synthase) is both necessary and sufficient to catalyze biosynthesis of geosmin from fernesyl diphosphate without requirement of any additional enzymes and redox cofactors, which solved the long-standing biosynthetic mystery.

Farnesol (3,7,11-trimethyl-2,6,10-dodecatrien-1-ol) is considered the universal precursor of the sesquiterpenes (Croteau 1987). Studies conducted by Dionigi et al. (1991) on the effect of farnesol on the growth and metabolism of the geosmin-producing actinomycete *Streptomyces tendeae* revealed that farnesol can inhibit geosmin synthesis, which in turn suppresses geosmin-producing species.

V Remediation of Off-Flavors

A *Conventional Physical Methods*

Management strategies for muddy and musty off-flavors are limited as geosmin and MIB are recalcitrant to conventional water treatment (Ho et al. 2007). However, some conventional physical techniques have been recommended. These sesquiterpenoids degrade over time and are purged from the fish, depending on their concentrations, water temperature, and water quality (Tucker and van der Ploeg 1999).

The minimum period required for fish to regain flavor quality is the cause of concern for aquaculturists (Dionigi et al. 2000). Lazur (2004) observed that holding fish in raceways with flow-through well water can purge geosmin and MIB off-flavors from fish; however, this process involves additional costs from harvesting and handling, tank facility overhead, water pumping, requirement of large amount of water (Johnsen and Dionigi 1993) and fish weight loss and mortality. Purging recirculating systems may be more practical but biological filters and other components of systems may become off-flavor sources (Johnsen and Dionigi 1993). van Breeman et al. (1991) reported an effective and environmentally friendly technique for the control T/O problems caused by algal activity in a reservoir, where sediment surface was disturbed with a harrow pulled by a boat.

Uptake and depuration of MIB from fish are important considerations in the design and implementation of systems to remove off-flavors from fish before processing (Johnsen et al. 1996). Flavor can be evaluated by tasting and assigning grades when fish is cooked in a microwave; flavor from a distinct to slight off-flavor is indicative of clearing of flavor, and the fish may soon be marketable (Lazur 2004). Song and O'Shea (2007) reported degradation of geosmin and MIB through ultrasonic irradiation, which may have potential applications in the removal of T/O compounds from potable water supplies and fish farms.

B Chemical Methods

Blue-green algae can be eliminated to some extent by chemical use in ponds (Wagner et al. 1999). One of the management practices to prevent or kill the growth of unwanted cyanobacteria includes the application of algicides to fish ponds (Tucker and van der Ploeg 1999; Lazur 2004). Copper sulfate, chelated-copper compounds, and diuron (3-[3,4-dichlorophenyl]-1,1-dimethylurea) are the USEPA-approved compounds for use in catfish production ponds as algicides (Schrader et al. 1998a,b, 2003; Schrader and Harries 2001; Tucker and Leard 1999). Most of the cyanobacteria are sensitive to 1–2 mg/L cupric ion, and some of them are affected even at 5 μg/L (Horne and Goldman 1974). An *Oscillatoria* species isolated in India was damaged at 1 mg/L copper sulfate after 8 d and failed to grow when transferred to growth medium (Arora and Gupta 1983). Studies conducted by Schrader and Blevins (2001) on the testing of trace elements revealed that copper had the most inhibitory effect on biomass and geosmin production at a concentration as low as 10.7 μM, and it was concluded that copper applied in the form of copper sulfate to the sediments of drained fish ponds might help prevent future off-flavor occurrences.

Prolonged use of copper sulfate can result in accumulation in the sediments, as shown in the Fairmont Lakes in Minnesota (Hanson and Stefan 1984) and Lake Monona, Wisconsin (Nichols et al. 1946). Large quantities of gesomin and MIB are retained in the blue-green cells, which may rupture on copper sulfate application with the result of rapid release of these intracellular odorous compounds (Negoro et al. 1988; Wu and Juttner 1988; Bowmer et al. 1992; Martin et al. 1991; Rosen et al. 1992; Utkilen and Frøshaug 1992). In Live Oak reservoir, Southern California, geosmin levels increased from 34 to 150 ng/L on copper sulfate application for control of geosmin-producing cyanobacteria.

The copper dose required to control a particular alga is not always effective because of its temporary effects (Izaguirre and Devall 1995) and the higher dose requirement, especially in alkaline waters, wherein it precipitates. Copper used for algal control has been found to be toxic to various freshwater fish, and speciation has a potential role in the toxicity of copper. It has been found that its continued use can result in copper-tolerant algal strains, requiring even higher doses for control (Izaguirre and Devall 1995), as evidenced in Lake Norrviken in Sweden (Ahlgren 1970), the Fairmont Lakes in Minnesota (Moyle 1949; Hanson and Stefan 1984), Mill Pond Reservoir in Massachusetts (McKnight et al. 1982), and Canadian prairie dugouts or farm ponds (Peterson and Swanson 1988). In Canada, continued application of copper sulfate favored the growth of *Oscillatoria* (Klassen et al. 1970). Copper tolerance has also been reported in various algae in lakes in Ontario, Canada (Stokes et al. 1973; Butler et al. 1980) and a river in England (Foster 1977).

Lyngbya, *Nostoc*, *and Phormidium* have been reported as copper-resistant blue-green algae (Palmer 1977). Izaguirre (1992) isolated a copper-tolerant benthic *Phormidium* sp., which produces MIB in Lake Mathews, California. The release of MIB in this reservoir has been linked with a cyanobacterium, *Oscillatoria curviceps*, first found in 1978 by Izaguirre et al. (1982). Later, by 1989, *Phormidium* had appeared all around the reservoir, following the decline of *O. curviceps*, which indicates that eradication of one taste-and-odor producer can be followed by the proliferation of another undesirable organism (Izaguirre 1992). The tolerance of *Phormidium* up to 3.5 mg/L copper in culture has been attributed to the repeated use of copper sulfate in the reservoir. Zimba et al. (2002) found that weekly applications of diuron to catfish ponds altered the taxonomic composition of the phytoplankton communities as the filamentous cyanobacteria were replaced by coccoid cyanobacteria. A copper-resistant strain of *M. aeruginosa* has been discovered by Garcia-Villada et al. (2004).

It has been observed (Izaguirre and Devall 1995; Tucker 2000; Han et al. 2001; Boylan 2001; Schrader et al. 2003; Tung et al. 2004) that synthetic algicides have the following adverse impacts: (i) toxicity toward phytoplankton that can lead to the death of the entire phytoplankton community and subsequent water quality deterioration; (ii) persistence in the environment; (iii) the public's negative perception of the use of synthetic herbicides in food fish production ponds; (iv) environmental safety issues from copper accumulation in the pond sediments; (v) adverse affect on microbial activity in pond sediments from long-term applications; (vi) deterioration of water quality resulting in the need for more aeration; (vii) pH fluctuation; (viii) dissolved oxygen depletion; and (ix) additional costs from multiple treatments as algae can reestablish in nutrient-rich water.

Studies conducted by Tung et al. (2004) on the effect of three different oxidants on MIB concentration in the presence of cyanobacteria in raw water revealed that ozonation was the most effective technique for the removal of both MIB and geosmin. Glaze et al. (1990) reported similar results in which 80%–90% of geosmin and MIB were removed by treatment with ozone. Ozonation appeared to affect the MIB concentrations by releasing it from damaged cells and oxidizing soluble MIB (Tung et al. 2004). Ozonation followed by biological filtration has the potential to provide effective treatment, as shown by Elhadi et al. (2004) in bench-scale experiments using granular activated carbon and sand for the removal of geosmin and MIB. Persson et al. (2007) used biofiltration to investigate differences between adsorption and biodegradation. They suppressed microbial activity by adding azide in granular activated carbon crushed in expanded clay. It was found that granular activated carbon still removed gesomin and MIB nearly unaffectedly, whereas in the clay biofilter, removal of both odorants ceased completely. Other oxidation processes using chlorine, chloramines, and potassium permanganate are ineffective for reducing geosmin and MIB as these oxidants cause only cell damage and the release of intracellular MIB into the water (Tung et al. 2004). These results are similar to those of Glaze et al. (1990). Peterson et al. (1995) also found that chlorine and permanganate caused extensive damage to algal cells, inducing the release of geosmin and dissolved organic carbon. Ashitani et al. (1988) observed an increase of MIB and geosmin concentrations in water following prechlorination at a water treatment plant. Jung et al. (2004) studied removal of geosmin and MIB by oxidation (O_3, Cl_2, ClO_2) and adsorption. They observed higher removal efficiency with increased ozone dosage and, in the case of pulverized activated carbon, adsorption efficiency of geosmin was superior to MIB. As an alternative to these synthetic algicides, natural compounds and extracts from plants are being screened for use in catfish aquaculture (Schrader et al. 2003; Meepagala et al. 2005).

C Environmentally Safe Plant-Derived Algicides

The discovery of eco-friendly, selective algicides that suppress the growth of the cyanobacteria implicated in musty off-flavor in pond-cultured catfish would be beneficial for the aquaculture industry. Green algae do not produce such undesirable odors, are good oxygenators of the water, and form a base for periphytic food growth in catfish production (Paerl and Tucker 1995); thus, the discovery of safe selective compounds that kill cyanobacteria would be beneficial for the aquaculture industry. Previous research (Schrader and Harries 2001; Schrader et al. 1998a,b) has identified several natural compounds that are selectively toxic toward *O. perornata*. 9,10-Anthraquinone, found in plant tannin extracts (Robinson 1967), has a high degree of selective toxicity toward *O. perornata* (Schrader et al. 1998a, b) and also inhibits its photosynthesis (Schrader et al. 2000). Previous studies shows that in comparison with copper-based products and diuron (half-life, 2 wk in pond water), anthraquinone-59 derived from the natural compound 9,10-anthraquinone has much lower persistence in pond water (half-life 19 hr) and also has greater selective toxicity toward cyanobacteria than other phytoplankton (Tucker 2000). In addition, the application of anthraquinone-59 in food fish production is advantageous in view of the public's negative perception of diuron.

Meepagala et al. (2005) extracted rutacridone epoxide from *Ruta graveolens* roots, which has potent selective algicidal activity toward the MIB-producing blue-green alga *Oscillatoria perornata*. Rutacridone epoxide is reported as a direct-acting mutagen, precluding its use as an agrochemical, and none of the synthetic analogues showed comparable activities to rutacridone epoxide (Meepagala et al. 2005). Many *Ruta* species are sources of diverse classes of natural products with biological activity including antifungal, phytotoxic, and antidotal (de Feo et al. 2002; Oliva et al. 2003). Oliva et al. (2003) demonstrated the presence of fungicidal constituents in the ethyl acetate extract of *Ruta graveolens* L. leaves against some agriculturally important fungi.

Tellez et al. (2001b) screened *F. cernua* extracts against two species of cyanobacteria and one species of green algae to determine their potential as a selective cyanobactericide. They found that the ether extract of *F. cernua* was selectively inhibitory against the cyanobacterium responsible for the MIB induced off-flavor associated with catfish farming operations.

D Lignocellulosic Agrowastes: Inexpensive Biosorbents

Activated carbon has been used very frequently for the removal of geosmin and MIB from natural water (Hung and Lin 2006), raw water (Cook et al. 2001), and drinking water (Hepplewhite et al. 2004; Elhadi et al. 2004; Liang et al. 2005). Nowack et al. (2004) investigated methods for tailoring a commercial, lignite-based granular activated carbon to enhance its adsorption of MIB from natural water. Cook et al. (2001) reported that powdered activated carbon (PAC) can effectively remove MIB and geosmin when the correct dose is applied, especially where a higher dose is required in the case of very turbid water. The high cost of activated carbon restricts its large-scale use for abatement of these metabolites, and in recent years the search for low-cost adsorbents has grown. By-products of lignocellulosic agroindustrial production have been studied for potential use as inexpensive biosorbents (Ng et al. 2002a, b). Barley straw inhibits the growth of cyanobacteria blooms (Barrett et al. 1996; Caffrey and Monahan 1999; Ferrier et al. 2005), which has been attributed to the antialgal activity of phenolics (tannins) present in the straw (Pillinger et al. 1994). Lignocellulosic materials have the advantage of being readily available because the world's industry utilizes less than 10% of raw material biomass from plantations (Pauli and Gravitis 1997). The remainder is waiting for effective utilization and could provide value-added products. Many other applications for these residues are in the process of being developed. Development of cost-effective and environment-friendly products from agricultural wastes/by-products and plantations for the aquatic bioremedition of brackishwater aquaculture is the objective of continued research of Central Institute of Brackishwater Aquaculture, Chennai, India (Krishnani et al. 2002, 2003, 2004, 2005, 2006b, 2006c; Krishnani and Ayyappan 2006; Parimala et al. 2004, 2007).

E Bioremediation

Microbes metabolize a broad range of aquatic pollutants by complex enzyme-catalyzed reactions. The genes encoding these proteins are localized on either large catabolic plasmids or the genomic DNA. Horizontal transfer of genes among bacteria has a major impact on the adaptability of bacteria during changing environmental conditions (Trevors et al. 1987). Gene bioaugmentation is the process of obtaining enhanced activity after gene transfer from an introduced donor organism into a member of the indigenous microbial environment (Pepper et al. 2002). This process has the potential to become a powerful tool in environmental management (DiGiovanni et al. 1996; Chen and Wilson 1997; Grommen and Verstraete 2002; Debashish et al. 2005).

To date, investigations related to bioremediation of geosmin and MIB are limited. Both are major causes of concern because they are difficult to remove by conventional water treatment methods (Lalezary et al. 1986). Biodegradation could be an alternative remediation technique, which needs to be investigated. Izaguirre et al. (1988) isolated mixed bacterial populations that biodegrade MIB slowly at ppb levels, whereas the related naturally occurring compound isoborneol was degraded rapidly, even at ppm levels, which may be attributed to the absence of the methyl group at carbon 2 in isoborneol and its presence in MIB, which might exert steric hindrance of the hydroxyl group (Medsker et al. 1969; Trudgill 1984).

Saadoun (2005) studied the ability of *Pseudomonas* sp. isolated from different soils contaminated with fuel spills to degrade MIB. The *Pseudomonas* group, especially *P. aeruginosa*, is common in freshwater and sediments (Hoadley 1968; Pellett et al. 1983) and well known for its metabolic versatility resulting from utilization of a wide range of substrates (Stainer et al. 1966). However, it has been reported that natural strains of this species do not have plasmids that encode degradative genes (Haas 1983). Walker and Higginbotham (2000) isolated an aquatic bacterium from pond water that could be a potential microbial algicide to lyse cells of some selected cyanobacteria, including species of *Anabaena* and *Oscillatoria*. Studies conducted by Klausen et al. (2005) showed that indigenous stream bacteria were capable of reducing the odors caused by geosmin and MIB produced by *Streptomyces*, and that enrichment with Luria-Bertani medium stimulated the degradation.

Lauderdale et al. (2004) isolated and characterized a bacterium implicated in aerobic degradation of MIB. Its 16S-rRNA phylogenetic analysis revealed that it is more closely related to *Bacillus fusiformis* and *Bacillus sphaericus*. Westerhoff et al. (2005) observed a magnitude change in MIB concentrations caused by thermal destratification of the reservoirs and concluded that biodegradation appeared more important than volatilization, photolysis, or adsorption. Saadoun (2005) modified the method of Jacobs et al. (1983) to determine the ability of different *Pseudomonas* sp. to degrade MIB-like compounds by transforming them to alcohol, detection of which would be an applicable approach for detecting the activity of microorganisms on this volatile compound.

Saadoun and El-Migdadi (1998) suggested that naturally occurring geosmin produced by *Streptomyces halstedii* could be degraded by specific species of gram-positive bacteria. They applied the technique of detection of alcohol production as a result of odorous compound oxidation for the screening of bacteria that degrade geosmin-like compounds. Hoefel et al. (2006) reported the cooperative degradation of geosmin by a consortium comprising three gram-negative bacteria isolated from a biologically active sand filter column, similar to cultured species such as *Sphingopyxis alaskensis*, *Novosphingobium stygiae*, and *Pseudomonas veronii*. They also observed that none of these three isolates was shown to be capable of degrading geosmin either individually or in any combination of two. Yagi et al. (1988) reported the degradation of more than 50% of geosmin and MIB adsorbed onto a bioactivated carbon filter seeded with *Bacillus subtilis*. Ho et al. (2007) reported biological sand filtration as an effective process for the biodegradation of MIB and geosmin, followed by batch bioreactor using biofilm. They identified a *Pseudomonas* sp., *Alphaproteobacterium*, *Sphingomonas* sp., and an *Acidobacteriaceae* member most likely involved in the biodegradation of geosmin.

Summary

Microbially derived off-flavors can adversely affect the beverage, food, water, and aquaculture industries. Off-flavor can temporarily be controlled by adopting best management practices such as proper aeration, liming, and dredging, and, more importantly, by avoidance of excessive nutrient use. Research studies focus on the effective means of control with the major emphasis on controlling the odor-causing algae populations and developing effective and selective algicides, which are not always available for use at the right time and can also have adverse impacts on the environment. Furthermore, selective application of synthetic algicides is not always recommended for reasons of inconsistency in the results and concerns regarding the frequent use of these chemicals, such as toxicity, accumulation of free copper, dissolved oxygen voids, increase in toxic ammonia and hydrogen sulfide, pH fluctuation, reduced photosynthetic activity, and reestablishment of algae in nutrient-rich water, thus requiring multiple treatments. Conversely, the plant-derived products appear to be environmentally safe and economical in view of their abundant availability and easy operational process. However, there needs to be more extensive work in this field. Precursors of sesquiterpene synthesis may selectively help to suppress off-flavor-producing species. Bioremedial measures by means of microbial degradation and gene bioaugmentation may be promising and are the subjects of much future research for effective controls.

Acknowledgments The authors are grateful to G. Izaguirre, Water Quality Section, Metropolitan Water District of Southern California, La Verne, California, for reviewing this manuscript extensively and for his valuable suggestions. The authors are grateful to Dr. David E. Cane, Department of Chemistry, Brown University, Rhode Island; Dr. K.K. Schrader, USDA, ARS, National Center for Natural Product Research, Mississippi; Dr. E. Vilalta and Dr. L. Blaha, Department of Ecology, University of Barcelona, Barcelona, Spain; Dr. S.B. Watson, National Water Research Institute, Burlington, Canada; Dr. I. Saadoun, Department of Biological Sciences, Jordon University of Science and Technology, Jordon; Dr. Tsair-Fuh Lin, Department of Environmental Engineering, National Cheng Kung University, Tainan City, Taiwan; Dr. Ingun Skjevrak, Norwegian Food Safety Authority, Sandnes, Norway; and Dr. H.H. Jelen, Institute of Food Technology, The August Cieszkowski Agricultural University of Poznan, Poland, for providing their valuable publications.

References

Ahlgren I (1970) Limnological studies of Lake Norrviken, a eutrophicated Swedish lake: II. Phytoplankton and its production. Schweiz Z Hydrol 32:353–396.

Anagnostidis K, Komárek J (1988) Modern approach to the classification system of cyanophytes: 3. Oscillatoriales. Arch Hydrobiol Suppl 80(1):327–472.

Arora A, Gupta AB (1983) The effect of copper sulfate on formation of separation discs in *Oscillatoria* species. Arch Hydrobiol 96:261–266.

Aschner M, Laventer C, Chorin-Kirsch I (1967) Off-flavor in carp from fishponds in the Coastal Plain and the Galil. Bamidgeh Bull Fish Cult Israel 19:23–25.

Ashitani K, Ishiba Y, Fujuwara K (1988) Behavior of musty odorous compounds during the process of water treatment. Water Sci Technol 20:261–267.

Baker PD (1992) Identification of Common Noxious Cyanobacteria: Part II. *Chroococcales, Oscillatoriales*. Research report 46. Australian Centre for Water Quality Research, Urban Water Research Association of Australia, Melbourne.

Baker PD, Cunliffe DA, Graham DK (1994) Biological Sources of Taste and Odour in the Millbrook-Mannum Water Supply System. Australian Centre for Water Treatment and Water Quality Research, Salisbury, South Australia.

Baker PD, Steffensen DA, Humpage AR, Nicholson BC, Falconer IR, Lanthois B, Fergusson KM, Saint CP (2001) Preliminary evidence of toxicity associated with the benthic cyanobacterium *Phormidium* in South Australia. Environ Toxicol 16:506–511.

Barrett PRF, Curnow J, Littlejohn JW (1996) The control of diatom and cyanobacterial blooms in reservoirs using barley straw. Hydrobiologia 340:307–311.

Bentley R, Meganathan R (1981) Geosmin and methylisoborneol biosynthesis in streptomycetes: evidence for an isoprenoid pathway and its absence in nondifferentiating isolates. FEBS Lett 125(2):220–222.

Berglind L, Holtan H, Skulberg OM (1983a) Case studies on off-flavours in some Norwegian lakes. Water Sci Technol 15(6/7):199–209.

Berglind L, Johnsen IJ, Ormerod K, Skulberg OM (1983b) *Oscillatoria brevis* (Kutz.) Gom. and some other especially odouriferous benthic cyanophytes in Norwegian inland waters. Water Sci Technol 15(6/7):241–246.

Blevins W (1980) Geosmin and other odourous metabolites of microbial origin. In: Gutherie FE, Perry JJ (eds) Introduction to Environmental Toxicology. Elsevier, New York, pp 350–357.

Bowmer KH, Padovan A, Oliver RL, Korth W, Ganf GG (1992) Physiology of geosmin production by *Anabaena circinalis* isolated from the Murrumbidgee River, Australia. Water Sci Technol 25(2):259–267.

Boylan JD (2001) Advancements in determining the role of barley straw as an algal control agent: notes. Iowa State University, Ames, IA.

Butler M, Haskew AEJ, Young MM (1980) Copper tolerance in the green algae *Chlorella vulgaris*. Plant Cell Environ 31:119–126.

Burlingame GA, Dann RM, Brock GL (1986) A case study of geosmin in Philadelphia water. J Am Water Works Assoc 78(3):56–61.

Burlingame GA, Muldowney JJ, Maddrey RE (1992) Cucumber flavor in Philadelphia's drinking water. J Am Water Works Assoc 84:92–97.

Caffrey JM, Monahan C (1999) Filamentous algal control using barley straw. Hydrobiologia 415:315–318.

Cane DE, Watt RM (2003) Expression and mechanistic analysis of a germacradienol synthase from *Streptomyces coelicolor* implicated in geosmin biosynthesis. Proc Natl Acad Sci USA 100:1547–1551.

Cane DE, He X, Kobayashi S, Omura S, Ikeda H (2006) Geosmin biosynthesis in *Streptomyces avermitilis*, molecular cloning, expression, and mechanistic study of the germacradienol/geosmin synthase. J Antibiot (Tokyo) 59(8):471–479.

Cees B, Zoeteman J, Piet GJ (1974) Cause and identification of taste and odour compounds in water. Sci Total Environ 3:103–115.
Chen S, Wilson DB (1997) Construction and characterization of *Escherichia coli* genetically engineered for bioremediation of Hg(2+)-contaminated environments. Appl Environ Microbiol 63:2442–2445.
Cook D, Newcombe G, Sztajnbok P (2001) The application of powdered activated carbon for 2-MIB and geosmin removal: predicting PAC doses in four raw waters. Water Res 35:1325–1333.
Cotsaris E, Bruchet J, Mallevialle J, Bursill DB (1995) The identification of odorous metabolites produced from algal monocultures. Water Sci Technol 31:251–258.
Croteau R (1987) Terpenoid natural products: a biosynthetic overview. In: Newman DW, Wilson KG (eds) Models in Plant Physiology and Biochemistry, vol II. CRC Press, Boca Raton, FL, pp 59–64.
Debashish G, Malay S, Barindra S, Joydeep M (2005) Marine enzymes. Adv Biochem Eng Biotechnol 96:189–218.
de Feo V, De Simone F, Senator F (2002) Potential allelochemicals from the essential oil of *Ruta graveolens*. Phytochemistry 61:573–578.
Desikachary TV (1959) Cyanophyta. Indian Council for Agricultural Research, New Delhi.
Dickschat JS, Martens T, Brinkhoff T, Simon M, Schulz S (2005) Volatiles released by a *Streptomyces* species isolated from the North Sea. Chem Biodivers 2(7):837–865.
DiGiovanni GD, Neilson JW, Pepper IL, Sinclair NA (1996) Gene transfer of *Alcaligenes eutrophus* JMP134 plasmid pJP4 to indigenous soil recipients. Appl Environ Microbiol 62: 2521–2526.
Dionigi CP, Millie DF, Johnson PB (1991) Effects of Farnesol and the off-flavor derivative geosmin on *Streptomyces tendae*. Appl Environ Microbiol 57:3429–3432.
Dionigi CP, Lawlor TE, McFarland GE, Johnson PB (1993) Evaluation of geosmin and 2-methylisoborneol on the histidine dependence of TA98 and TA 100 *Salmonella typhimurium* tester strains. Water Res 27:1615–1618.
Dionigi CP, Bett K, Johnsen P, McGillberry J, Millie D, Vinyard B (1998) Variation in channel catfish *Ictalurus punctatus* flavor quality and its quality control implications. J World Aquacult Soc 29:140–154.
Dionigi CP, Johnson PB, Vinyard BT (2000) The recovery of flavor quality by channel catfish. N Am J Aquacult 62(3):189–194.
Elhadi SL, Huck PM, Slawson RM (2004) Removal of geosmin and 2-methylisoborneol by biological filtration. Water Sci Technol 49(9):273–280.
Engle CR, Pounds GL, van der Ploeg M (1995) The cost of off-flavor. J World Aquacult Soc 26:297–306.
Eynard F, Mez K, Walther J (2000) Risk of cyanobacterial toxins in Riga waters (Latvia). Water Res 34:2979–2988.
Farmer LJ, McConnell JM, Hagan TD, Harper DB (1995) Flavor and off flavor in wild and farmed Atlantic salmon from locations around Northern Ireland. Water Sci Technol 31:259–264.
Ferrier MD, Butler BR, Terlizzi DE, Lacouture RV (2005) The effects of barley straw (*Hordeum vulgare*) on the growth of freshwater algae. Bioresour Technol 96(16):1788–1795.
Foster PL (1977) Copper exclusion as a mechanism of heavy metal tolerance in a green algae. Nature (Lond) 269:322–323.
Frank HK (1977) Occurrence of patulin in fruit and vegetables. Ann Nutr Aliment 4:459–465.
Fravel DR, Connick WJ Jr, Grimm CC, Lloyd SW (2002) Volatile compounds emitted by sclerotia of *Sclerotinia minor*, *Sclerotinia sclerotiorum*, and *Sclerotium rolfsii*. J Agric Food Chem 50:3761–3764.
Frisvad JC, Samson RA, Rassing BR, van der Horst MI, van Rijn FT, Stark J (1997) *Penicillium discolor*, a new species from cheese, nuts and vegetables. Antonie Van Leeuwenhoek 72:119–126.
Gagne E, Ridal J, Blaise C, Brownlee B (1999) Toxicological effects of geosmin and 2-methylisoborneol on rainbow trout hepatocytes. Bull Environ Contam Toxicol 63:174–180.

Garcia-Villada L, Rico M, Altamirano M, Sanchez-Martin L, Lopez-Rodas V, Costas E (2004) Occurrence of copper resistant mutants in the toxic cyanobacteria *Microcystis aeruginosa:* characterization and future implications in the use of copper sulfate as algicide. Water Res 38:2207–2213.

Gerber NN (1968) Geosmin from micro-organisms is *trans*-1,10-dimethyl-*trans*-9-decalol. Tetrahedron Lett 25:2971–2974.

Gerber NN (1969) A volatile metabolite of actinomycetes: 2-methyl-isoborneol. J Antibiot (Tokyo) 22:508–509.

Gerber NN (1977) Three highly odorous metabolites from an actinomycete: 2-isopropyl-3-methoxy-pyrazine, methylisoborneol, and geosmin. J Chem Ecol 3:475–482.

Gerber NN (1979) Volatile substances from actinomycetes: their role in the odor pollution of water. Crit Rev Microbiol 7:191–214.

Gerber NN (1983) Volatile substances from actinomycetes: their role in the odour pollution of water. Water Sci Technol 15:115–125.

Gerber NN, Lechevalier HA (1965) Geosmin, an earthy smelling substance isolated from actinomycetes. Appl Microbiol 13:935–938.

Glaze WH, Schep R, Chauncey W, Ruth EC, Zarnoch JJ, Aieta EM, Tate CH, McGuire MJ (1990) Evaluating oxidants for the removal of model taste and odor compounds from a municipal water supply. J Am Water Works Assoc A 82:79–84.

Grommen R, Verstraete W (2002) Environmental biotechnology: the ongoing quest J Biotechnol 98:113–123.

Gust B, Challis GL, Fowler K, Kieser T, Charter KF (2003). PCR-targeted Streptomyces gene replacement identifies a protein domain needed for biosynthesis of the sesquiterpene soil odor geosmin. Proc Natl Acad Sci U S A 100(4):1541–1546.

Han FX, Hargreaves JA, Kingery WL, Hugget DB, Schlenk DK (2001) Accumulation, distribution, and toxicity of copper sulfate in sediments of catfish ponds receiving periodic copper sulfate applications. J Environ Qual 30:912–919.

Hanson MJ, Stefan HG (1984) Side effects of 58 years of copper sulfate treatment of the Fairmont Lakes, Minnesota. Water Resour Bull 20:889–900.

Harborne JB, Tmoas-Barberan EA (1989) Ecological Chemistry and Biochemistry of Plant Terpenoids. Clarendon Press, Oxford.

Harrawijn P, Oosten van AM, Piron PG (2001) Natural Terpenoids as Messengers. Kluwer, Dordrecht.

Haas D (1983) Genetic aspects of biodegradation by pseudomonads. Experientia (Basel) 39:1199–1213.

Hayes PK, Burch MD (1989) Odorous compounds associated with algal blooms in south Australian waters. Water Res 23:115–121.

He X, Cane DE (2004) Mechanism and stereochemistry of the germacradienol/germacrene D synthase of *Streptomyces coelicolor* Ae(2). J Am Chem Soc 126(9): 2678–2679.

Henley DE (1970) Odorous metabolite and other selected studies of Cyanophyta. Doctoral dissertation. North Texas State University, Denton, TX.

Hepplewhite C, Newcombe G, Knappe DR (2004) NOM and MIB: who wins in the competition for activated carbon adsorption sites? Water Sci Technol 49:257–265.

Ho L, Hoefel D, Bock F, Saint CP, Newcombe G (2007) Biodegradation rates of 2-methylisoborneol (MIB) and geosmin through sand filters and in bioreactors. Chemosphere. 66(11):2210–2218

Hoadley AW (1968) On the significance of *Pseudomonas aeruginosa* in surface waters. JN Engl Water Works Assoc 82:99–111.

Hoefel D, Ho L, Aunkofer W, Monis PT, Keegan A, Newcombe G, Saint CP (2006) Cooperative biodegradation of geosmin by a consortium comprising three gram negative bacteria isolated from the biofilm of a sand filter. Lett Appl Microbiol 3(4):417–423.

Horne AJ, Goldman CR (1974) Suppression of nitrogen fixation by blue green algae in a Eutrophic Lake With trace additions of copper. Science 83:409–411.

Hosaka M, Murata K, Iikura Y, Oshimi A, Udagawa T (1995) Off-flavor problem in drinking water of Tokyo arising from the occurrence of musty odor in a downstream tributary. Water Sci Technol 31(11):29–34.

Hu TL, Chiang PC (1996) Odorous compounds from a cyanobacterium in a water purification plant in central Taiwan. Water Res 30(10):2522–2525.

Hung HW, Lin TF (2006) Predicting the adsorption capacity and isotherm curvature of organic compounds onto activated carbons in natural waters. Environ Technol 27:255–267.

Hsieh T, Tanchotikul U, Mathella E (1988) Identification of geosmin as the major muddy off-flavor of Louisiana brackishwater clam Rangia cuneata. J Food Sci 53:1228–1229.

Izaguirre G (1992) A copper-tolerant *Phormidium* species from Lake Mathews, California, that produces 2-methylisoborneol and geosmin. Water Sci Technol 25(2):217–224.

Izaguirre G, Devall (1995) Resource control for management of taste and odor problems. In: Suffet IH, Mallevialle J, Kawczynski E (eds) Advances in Tastes-and–Odor Treatment and Control. American Water Works Association Research Foundation, Denver, pp 37–43.

Izaguirre G, Taylor WD (1995) Geosmin and 2-methylisoborneol production in a major aqueduct system. Water Sci Technol 31:41–48.

Izaguirre G, Taylor WD (1998) A *Pseudanabaena* species from Castaic Lake, California, that produces 2-methylisoborneol. Water Res 32(5):1673–1677.

Izaguirre G, Taylor WD (2004) A guide to geosmin- and MIB-producing cyanobacteria in the United States. Water Sci Technol 49:19–24.

Izaguirre G, Hwang CJ, Krasner SW, McGuire MJ (1982) Geosmin and 2-methylisoborneol from cyanobacteria in three water supply systems. Appl Environ Microbiol 43:708–714.

Izaguirre G, Hwang CJ, Krasner SW, Mcguire MJ (1983) Production of 2-methylisoborneol by two benthic cyanophyta. Water Sci Technol 15:211–220.

Izaguirre G, Hwang CJ, Krasner SW (1984) Investigations into the source of 2-methylisoborneol in Lake Perris, California. In: Proceedings of 11th Annual Water Quality Technology Conference, 4–7 December, 1983, Norfolk VA, USA. American Water Works Association, Denver, CO.

Izaguirre G, Wolfe RL, Means EG (1988) Degradation of 2-methylisoborneol by aquatic bacteria. Appl Environ Microbiol 54:2424–2431.

Izaguirre G, Taylor WD, Pasek J (1999) Off-flavor problems in two reservoirs, associated with planktonic *Pseudanabaena* species. Water Sci Technol 40(6):85–90.

Izaguirre G, Jungblut A-D, Neilan BA (2007) Benthic cyanobacteria (Oscillatoriaceae) that produce microcystin-LR, isolated from four reservoirs in southern California. Water Res 41:492–498.

Jacobs CJ, Prior BA, De Kock MJ (1983) A rapid screening method to detect ethanol production by microorganisms. J Microbiol Methods 1:339–342.

Jardine C, Gibson N, Hrudey S (1999) Detection of odour and health risk perception of drinking water. Water Sci Technol 40:91–98.

Jelen HH, Malgorzata M, Zawirska-Wojtasiak R, Malgorzata W, Wasowicz E (2003) Determination of geosmin, 2-methylisoborneol, and a musty-earthy odor in wheat grain by SPME-GC-MS, profiling volatiles, and sensory analysis. J Agric Food Chem 51:7079–7085.

Jiang J, He X, Cane DE (2006) Geosmin biosynthesis. *Streptomyces coelicolor* germacradienol / germacrene D synthase converts farnesyl diphosphate to geosmin. J Am Chem Soc 128:8128–8129.

Johnsen PB, Dionigi CP (1993) Physiological approaches to the management of off-flavors in farm-raised channel catfish. J Appl Aquacult 3:141–161.

Johnsen PB, Lloyd SW, Vinyardm BT, Dionigi CP (1996) Effect of temperature on the uptake and depuration of 2-methylisoborneol (MIB) in channel catfish *Ictalurus punctatus*. J World Aquacult Soc 27:15–20.

Jung SW, Baek KH, Yu MJ (2004) Treatment of taste and odour material by oxidation and adsorption. Water Sci Technol 49(9):289–295.

Jüttner F (1984) Dynamics of the volatile organic substances associated with cyanobacteria and algae in a eutrophic shallow lake. Appl Environ Microbiol 47:814–820.

Jüttner E, Hoflacher B, Wurster K (1986) Seasonal analysis of volatile organic biogenic substances (VOBS) in freshwater phytoplankton populations dominated by *Dinobryon*, *Microcystis* and *Aphanizomenon*. J Phycol 22:169–175.

Jüttner F (1995) Physiology and biochemistry of odorous compounds from freshwater cyanobacteria and algae. Water Sci Technol 31(11):69–78.

Kajino M, Sakamoto K (1995) The relationship between musty odor causing organisms and water quality in Lake Biwa. Water Sci Technol 31:153–158.

Karahadian C, Josephson DB, Lindsay RC (1985) Volatile compounds from *Penicillium* sp. contributing musty earthy notes to Brie and Camembert cheese flavors. J Agric Food Chem 33:339–343.

Kikuchi T, Mimura T, Harimaya K, Yano H, Arimoto T, Masada Y, Inoue T (1973) Odorous metabolite of blue-green alga: *Schizothrix muelleri* NAGELI collected in the southern basin of lake Biwa. Identification of geosmin. Chem Pharm Bull 21:2342–2343.

Kikuchi T, Kadota S, Suehara H, Nishi A, Tsubaki K (1981) Odorous metabolites of a fungus, *Chaetomium globosum* Kinze ex Fr. Identification of geosmin, musty-smelling compound. Chem Pharm Bull 29:1781–1784.

Kilkast D (1993) Sensory evaluation of taints and off-flavors. In: Saxby MJ (ed) Food Taints and Off-Flavors. Chapman & Hall, Glasgow, Scotland, pp 1–32.

Klassen EH, Rempel G, Pick AR (1970) Pilot studies for quality management in terminal reservoir. J Am Water Works Assoc 62(9):555–560.

Klausen C, Nicolaisen MH, Strobel BW, Warnecke F, Nielsen JL, Jorgensen NO (2005) Abundance of actinobacteria and production of geosmin and 2-methylisoborneol in Danish streams and fish ponds. FEMS Microbiol Ecol 52:265–278.

Komarkova-Legnerova J, Cronberg G (1992) New and recombined filamentous cyanophytes from 6 lakes in South Scania, Sweden. Algol Stud 67:21–32.

Krasner SW, McGuire MJ, Ferguson VB (1985) Tastes and odors: the flavor profile method. J Am Water Works Assoc 77(3):34–39.

Krishnani KK, Ayyappan S (2006) Heavy metals remediation of water using plant & lignocellulosic agrowastes. Rev Environ Contam Toxicol 88: 64–85.

Krishnani KK, Gupta BP, Joseph KO, Muralidhar M, Nagavel A (2002) Studies on the use of neem products for removal of ammonia from brackishwater. J Environ Sci Health A37:893–904.

Krishnani KK, Gupta BP, Joseph KO, Muralidhar M, Sarda C, Nagavel A, Parimala V (2003) Decontamination of nitrogenous toxicants from brackishwater using natural plant and animal extracts. Bull Environ Contam Toxicol 71:196–203.

Krishnani KK, Parimala V, Meng X (2004) Detoxification of hexavalent chromium from coastal water using lignocellulosic waste. Water S Afr 30:541–545.

Krishnani KK, Gupta BP, Pillai SM, Ravichandran P (2005) Environmental challenges and management in coastal aquaculture. In: Mohan JM, Sheela PJ (eds) Fisheries and Environment—Matsyagandha. CMFRI (ICAR), Cochin, India, June 2006, 91(6):69–74 (in Hindi).

Krishnani KK, Ravichandran P, Gupta BP (2006a) Pre-harvest off flavor in aquaculture. In: Mohan JM, Sheela PJ (eds) Proceedings of National Conference on Livelihood Issues in Fisheries & Aquaculture, CMFRI (ICAR), Cochin, India, 90:45–46 (in Hindi).

Krishnani KK, Parimala V, Gupta BP, Azad IS, Meng X, Abraham M (2006b) Bagasse assisted bioremediation of shrimp farm wastewater. Water Environ Res 78:938–950.

Krishnani KK, Parimala V, Gupta BP, Azad IS, Shekhar MS (2006c) Bioremediation of nitrite from brackishwater using lignocellulosic waste—bagasse. Asian Fish Sci 19(4):429–444.

La Guerche S, Chamount S, Blancard D, Dubourdieu D, Darriet P (2005) Origin of (−)-geosmin on grapes: on the complementary action of two fungi, *Botrytis cinerea* and *Penicillium expansum*. Antonie Van Leeuwenhoek 88:131–139.

Lalezary S, Pirbazari M, McGuire MJ (1986) Oxidation of five earthy-musty taste and odor compounds. J AM Water Works Assoc 78:62–69.

Lauderdale CV, Aldrich HC, Lindner AS (2004) Isolation and characterization of a bacterium capable of removing taste- and odor-causing 2-methylisoborneol from water. Water Res 38(19):4135–4142.

Lazur A (2004) Off flavor tastes in aquaculture. Maryland Aquafarmer 3:1–4.

Leventer H, Eren J (1970) Taste and odour in the reservoirs of the Israel national water system. In: Shuval HI (ed) Developments in Water Quality Research. Humphrey Science, Ann Arbor, MI, pp 19–37.

Liang C, Wang D, Yang M, Sun W, Zhang S (2005) Removal of earthy-musty odorants in drinking water by powdered activated carbon. J Environ Sci Health A40:767–778.

Lin TF, Wong JY, Kao HP (2002) Correlation of musty odor and 2-MIB in two drinking water treatment plants in south Taiwan. Sci Total Environ 289:225–235.

Lovell R, Broce D (1985) Cause of musty flavor in pond-cultured penaeid shrimp. Aquaculture 50:169–174.

Lu G, Fellmann JK, Edwards CG, Mattinson DS, Navazio J (2003) Quantitative determination of geosmin in red beets (*Beta vulgaris* L.) using headspace solid phase microextraction. J Agric Food Chem 51:1021–1025.

Lundgren BV, Boren H, Grimvall A, Savenhed R, Wigilus B (1988) The efficiency and relevance of different concentration methods for the analysis of off-flavors in water. Water Sci Technol 20:81.

Maga JA (1987) Musty/earthy aromas. Food Rev Int 3:269–284.

Martin JF, McCoy CP, Tucker CS, Bennett L (1988) 2-Methylisoborneol implicated as a cause of off-flavor in channel catfish *Ictalurus punctatus* (Rafinesque) from commercial culture ponds in Mississippi. Aquacult Fish Manag 19:151–157.

Martin JF, Izaguirre G, Waterstrat P (1991) A planktonic *Oscillatoria* species from Mississippi catfish ponds that produces the off-flavor compound 2-methylisoborneol. Water Res 25:1447–1451.

Matsumoto A, Tsuchiya Y (1988) Earthy-musty odor-producing cyanophytes isolated from five water areas in Tokyo. Water Sci Technol 20(8/9):179–183.

McGuire M (1995) Off-flavor as the consumer's measure of drinking water safety. Water Sci Technol 31:1–8.

McKnight DM, Chisholm SW, Morel FMM (1982) Copper sulfate treatment of lakes and reservoirs: chemical and biological considerations. Tech. note no. 24. Ralph M. Parsons Laboratory, Massachusetts Institute of Technology, Cambridge.

Medsker LL, Jenkins D, Thomas JF (1968) Odorous compounds in natural waters. An earthy-smelling compound associated with blue-green algae and actinomycetes. Environ Sci Technol 2:461–464.

Medsker LL, Jenkins D, Thomas JF, Koch C (1969) Odorous compounds in natural waters: 2-exo-hydroxy-2-methylbornane, the major odorous compound produced by several actinomycetes. Environ Sci Technol 3:476–477.

Meepagala KM, Schrader KK, Wedge DE, Duke SO (2005) Algicidal and antifungal compounds from the roots of *Ruta graveolens* and synthesis of their analogs. Phytochemistry 66:2689–2695.

Mercier J, Jimenez JI (2004) Control of fungal decay of apples and peaches by the biofumigant fungus *Muscodor albus*. Postharvest Biotechnol 31:1–8.

Mohren S, Jüttner F (1983) Odorous compounds of different strains of *Anabaena* and *Nostoc* (cyanobacteria). Water Sci Technol 15:221–228.

Motohiro T (1983) Tainted fish caused by petroleum compounds—a review. Water Sci Technol 15(6/7):75–83.

Moyle JB (1949) The use of copper sulfate for algal control an its biological implications. In: Limnological Aspects of Water Supply and Waste Disposal. American Association for the Advancement of sciences, Washington DC.

Naes H, Utkilen H, Post A (1988), Factors influencing geosmin production by the cyanobacterium *Oscillatoria brevis*. Water Sci Technol 20:125–131.

Nakajima MT, Ogura Y, Kusama N, Iwabuchi T, Imawaka A, Araki T, Sasaki T, Hirose E (1996) Inhibitory effects of odour substances, geosmin and 2-methylisoborneol, on early development of sea urchins. Water Res 30:2508–2516.

Negoro T, Ando M, Ichikawa N (1988) Blue-green algae in Lake Biwa which produce earthy-musty odors. Water Sci Technol 20(8/9):117–123.
Negri AP, Jones GJ, Hindmarsh M (1995) Sheep mortality associated with paralytic shellfish poisoning toxins from the cyanobacterium *Anabaena circinalis*. Toxicon 33:1321–1329.
Nichols MS, Henkel T, McNall D (1946) Copper in lake muds of the Madison Area. Trans Wisc Acad Sci Arts Lett 38:333–350.
Nielsen JL, Klausen C, Nielsen PH, Burgord M, Jorgensen NO (2006) Detection of activity among uncultured Actinobacteria in a drinking water reservoir. FEMS Microbiol Ecol 55: 432–438.
Ng C, Losso JN, Marshall WE, Rao RM (2002a) Physical and chemical properties of selected agricultural by-product-based activated carbons and their ability to adsorb geosmin. Bioresour Technol 84:177–185.
Ng C, Losso JN, Marshall WE, Rao RM (2002b) Freundlich adsorption isotherms of agricultural by-product-based powdered activated carbons in a geosmin-water system. Bioresour Technol 85:131–135.
Nowack KO, Cannon FS, Mazyck DW (2004) Enhancing activated carbon adsorption of 2-methylisoborneol: methane and steam treatments. Environ Sci Technol 38:276–284.
Nunes C, Usall J, Teixido N, Torres R, Vinas I (2002) Control of *Penicillium expansum* and *Botrytis cinerea* on apples and pears with the combination of *Candida sake* and *Pantoea agglomerans*. J Food Protect 65:178–184.
Oliva A, Meepagala KM, Wdge DE, Harries D, Hale AL, Aliotta G, Duke SO (2003) Natural fungicides of from *Ruta graveolens* L. leaves, including a new quinolene alkaloid. J Agric Food Chem 51:890–896.
Palmer CM (1977) Algae and water pollution. EPA-600/9-77-036. Municipal Environmental Research Laboratory, U.S. Environmental Protection Agency, Cincinnati, OH.
Paerl HW, Tucker CS (1995) Ecology of blue-green algae in aquaculture ponds. J World Aquacult Soc 26:109.
Parimala V, Krishnani KK, Gupta BP, Jayanthi M, Abraham M (2004) Phytoremediation of chromium from seawater using five different products from coconut husk. Bull Environ Contam Toxicol 73:31–37.
Parimala V, Krishnani KK, Gupta BP, Ragunathan R, Pillai SM, Ravichandran P (2007) Removal of ammonia and nitrite from coastal water using low cost agricultural waste. Bull Environ Contam Toxicol 78(3–4):288–293.
Pauli G, Gravitis J (1997) Environmental Management of Plantations: Through Zero Emission Approach-Plantation management for the 21st Century. Proceedings of the International Planters Conference on Plantation Management for the 21st Century, Kuala Lumpur, Malaysia, Vol 1. The Incorporated Society of Planters, pp 193–207.
Pellett S, Bigley DV, Grimes DJ (1983) Distribution of *Pseudomonas aeruginosa* in a riverine ecosystem. Appl Environ Microbiol 45:328–332.
Pepper IL, Gentry TJ, Newby DT, Roane TM, Josephson KL. (2002) The role of cell bioaugmentation and gene bioaugmentation in the remediation of co-contaminated soils. Environ Health Perspect 110:943–946.
Persson PE (1979) Notes on muddy odour. III. Variability of sensory response to 2-methylisoborneol. Aqua Fenn 9:48–52.
Persson PE (1980) Muddy odour in fish from hypertrophic waters. Dev Hydrobiol 2:203–208.
Persson PE (1981) The etiology of muddy odour in fish and water. Finn Fish Res 4:1–13.
Persson PE (1983) Off-flavors in aquatic ecosystems: an introduction. Water Sci Technol 15(6/7):1–11.
Persson PE (1988) Odorous algal cultures in culture collections. Water Sci Technol 20:211–213.
Persson F, Heinicke G, Hedberg T, Hermansson M, Uhl W (2007) Removal of geosmin and MIB by biofiltration: an investigation discriminating between adsorption and biodegradation. Environ Technol 28(1):95–104.
Peterson HG, Swanson SM (1988) Physical, chemical and biological warfare against algae. In: Proceedings, 40th Annual Convention of the Western Canadian Water and Wastewater Association, Winnipeg, Manitoba, Canada, pp 153–161.

Peterson HG, Hrudey SE, Cantin IA, Perley TR, Kenefrick SL (1995) Physiological toxicity, cell membrane damage and the release of dissolved organic carbon and geosmin by *Aphanizomenon flos-aquae* after exposure to water treatment chemicals. Water Res 29:1515–1523.

Piet GJ, Zoeteman BCJ, Kraayeveld AJA (1972) Earthy smelling substances in surface waters of the Netherlands. Water Treat Exam 21:281.

Pillinger JM, Cooper JA, Ridge I (1994) Role of phenolic compounds in the antialgal activity of barley straw. J Chem Ecol 20:1557–1569.

Polak E, Provasi J (1992) Odour sensitivity to geosmin enantiomers. Chem Senses 17:23–26.

Rashash DMC, Hoehn RC, Dietrich AM, Grizzard TJ, Parker BC (1996) Identification and Control of Odorous Algal Metabolites. AWWA Research Foundation, Denver, CO.

Robinson T (1967) The Constituents of Higher Plants. Burgess, Minneapolis, MN.

Rosen AA, Mashni CI, Safferman RS (1970) Recent developments in the chemistry of odour in water: the cause of earthy-musty odour. Water Treat Exam 19:106.

Rosen BH, Macleod BW, Simpson MR (1992) Accumulation and release of geosmin during growth phase of *Anabaena circinalis* (Kutz) Rabenhorst. Water Sci Technol 25(2): 185–190.

Saadoun I (2005) Production of 2-methylisoborneol by *Streptomyces violaceusniger* and its transformation by selected species of *Pseudomonas*. J Basic Microbiol 45:236–242.

Saadoun I, El-Migdadi F (1998) Degradation of geosmin- like compounds by selected species of gram-positive bacteria. Lett Appl Microbiol 28:98–100.

Saadoun I, Schrader KK, Blevins WT (1997) Identification of 2-methylisoborneol (MIB) and geosmin as volatile metabolites of *Streptomyces violaceusniger*. Actinomycetes 8:37–41.

Saadoun IMK, Schrader KK, Blevins WT (2001) Environmental and nutritional factors affecting geosmin synthesis by *Anabaena* sp. Water Res 35(5):1209–1218.

Sabater S, Vilalta E, Gaudes A, Guasch H, Muñoz I, Romani A (2003) Ecological implications of mass growth of benthic cyanobacteria in rivers. Aquat Microb Ecol 32:175–184.

Saxby MJ (1993) A survey of chemicals causing taints and off-flavors in foods. In: Saxby M (ed) Food Taints and Off-Flavors. Chapman & Hall; Glasgow, pp 35–62.

Schnurer J, Olssen J, Borjesson T (1999) Fungal volatiles as indicators of food and feeds spoilage, Fung Genet Biol 27:209–217.

Scholler CEG, Gurtler H, Pedersen R, Molin S, Wilkins K (2002) Volatile metabolites from Actinomycetes. J Agric Food Chem 50:2615–1621.

Schrader K, Blevins W (1993) Geosmin producing species of *Streptomyces* and *Lyngbya* from aquaculture ponds. Can J Microbiol 39:834–840.

Schrader KK, Blevins WT (2001) Effects of carbon source, phosphorus concentration, and several micronutrients on biomass and geosmin production by *Streptomyces halstedii*. Ind Microbiol Biotechnol 26:241–247.

Schrader KK, Dennis ME (2005) Cyanobacteria and earthy/musty compounds found in commercial catfish (*Ictalurus punctatus*) ponds in the Mississippi Delta and Mississippi-Alabama Blackland Prairie. Water Res 39:2807–2814.

Schrader KK, Harries MD (2001) Compounds with selective toxicity towards the must-odour cyanobacterium *Oscillatoria perornata*. Bull Environ Contam Toxicol 66:801–807.

Schrader KK, de Regt MQ, Tidwell PD, Tucker CS, Duke SO (1998a) Compounds with selective toxicity towards the off-flavor metabolite-producing cyanobacterium *Oscillatoria* cf. *chalybea*. Aquaculture 163:85–99.

Schrader KK, de Regt MQ, Tidwell PR, Tucker CS, Duke SO (1998b) Selective growth inhibition of the musty-odour producing cyanobacterium *Oscillatoria* cf. *chalybea* by natural compounds. Bull Environ Contam Toxicol 60:651–658.

Schrader KK, Dayan FE, Allen SN, de Regt MQ, Tucker CS, Paul RN Jr (2000) 9,10-Anthraquinone reduces the photosynthetic efficiency of *Oscillatoria perornata* and modifies cellular inclusions. Int J Plant Sci 161:265–270.

Schrader KK, Nanayakkara NP, Tucker CS, Rimando AM, Ganzera M, Schaneberg BT (2003) Novel derivatives of 9,10-anthraquinone are selective algicides against the musty-odor cyanobacterium *Oscillatoria perornata*. Appl Environ Microbiol 69:5319–5327.

Seligman K, Enos AK, Lai HH (1992) A comparison of 1988–1990 flavor profile analysis results with water conditions in two northern California reservoirs. Water Sci Technol 25:19–25.

Siegmund B, Pollinger-Zierler B (2006) Odor threshold of microbially induced off-flavor compounds in apple juice. J Agric Food Chem 54(16):5984–5989.

Silvey JKG, Russell JC, Redden DR, McCormick WC (1950) Actinomycetes and common tastes and odors. J Am Water Works Assoc 42:1018–1026.

Sivonen K (1982) Factors influencing odour production by actinomycetes. Hydrobiologia 86:165–170.

Skjevrak I, Lund V, Ormerod K, Herikstad H (2005) Volatile organic compounds in natural biofilm in polythene pipes supplied with lake water and treated water from the distribution network. Water Res 39:4133–4141.

Slater GP, Blok VC (1983a) Volatile compounds of the Cyanophyceae: a review. Water Sci Technol 15:181–190.

Slater GP, Blok VC (1983b) Isolation and identification of odorous compounds from a lake subject to cyanobacterial blooms. Water Sci Technol 15:229–240.

Song W, O'Shea KE (2007) Ultrasonically induced degradation of 2-methylisoborneol and geosmin. Water Res 41(12):2672–2678.

Spiteller D, Jux A, Piel J, Boland W (2002) Feeding of [5,5-2H(2)]-1-desoxy-D-xylulose and [4,4,6,6,6-2*H*(5)]-mevalolactone to a geosmin-producing *Streptomyces* sp. and *Fossombronia pusilla*. Phytochemistry 61:827–834.

Stainer RY, Palleroni NJ, Douderoff M (1966) The aerobic pseudomonads: a taxonomic study. J Gen Microbiol 43:159–271.

Stokes PM, Hutchinson TC, Kranter K (1973) Heavy metal tolerance algae isolated from contaminated lakes near Sudbury. Ontario Can J Bot 51:2155–2168.

Suffet IH, Corado A, Chou D, Butterwoth S, McGuire MJ (1996) AWWA survey of taste and odor. J Am Water Works Assoc 88:168–180.

Sugiura N, Yagi O, Sudo R (1986) Musty odor from blue-green alga. *Phormidium tenue* in Lake Kasumigaura. Environ Technol Lett 7:77–86.

Sugiura N, Nishimura O, Inamori Y, Ouchiyama T, Sudo R (1997) Grazing characteristics of musty odour compound producing *Phormidium tenue* by a microflagellate, *Monas guttula*. Water Res 31:2792–2796.

Sugiura N, Iwami N, Inamori Y, Nishimura O, Sudo R (1998) Significance of attached cyanobacteria relevant to the occurrence of musty odor in Lake Kasumigaura. Water Res 32:3549–3554.

Taylor WD, Losee RF, Torobin M, Izaguirre G, Sass D, Khiari D, Atasi K (2006) Early warning and management of surface water taste-and-odor events. In: Geosmin- and MIB-Producing Cyanobacteria Found in the United States. American Water Works Association Research Foundation, Denver, CO, chapter 4.

Tabachek JL, Yurkowski M (1976) Isolation and identification of blue green algae producing muddy odor metabolites, geosmin, and 2-methylisoborneol, in saline lakes in Manitoba. J Fish Res Board Can 33:25–35.

Tellez MR, Schrader KK, Kobaisy M (2001a) Volatile components of the cyanobacterium *Oscillatoria perornata* (Skuja). J Agric Food Chem 49:5989–5992.

Tellez MR, Estell R, Fredrickson Ed, Powell J, Wedge D, Schrader K, Kobaisy M (2001b) Extracts of *Flourensia cernua* (L): volatile constituents and antifungal, and antitermite bioactivities. J Chem Ecol 27:2263–2273.

Trevors JT, Barkay T, Bourquin AW (1987) Gene transfer among bacteria in soil and aquatic environments: a review. Can J Microbiol 33:191–195.

Trudgill PW (1984) Microbial degradation of the alicyclic ring. In: Gibson DT (ed) Microbial Degradation of Organic Compounds. Dekker, New York.

Tsuchiya Y, Matsumuto A (1988) Identification of volatile metabolites produced by blue-green algae. Water Sci Technol 20:149–155.

Tsuchiya Y, Matsumoto A, Okamoto T (1981) Identification of volatile metabolites produced by blue-green algae, *Oscillatoria splendida, O. amoena, O. geminata*, and *Aphanizomenon* sp. Yakugaku Zasshi 101:852. (in Japanese).

Tucker CS (2000) Off-flavor problems in aquaculture. Rev Fish Sci 8:45–88.
Tucker CS, Leard AT (1999) Managing catfish off-flavors with diuron. Fact sheet 003–revised. Thad Cochran National Warmwater Aquaculture Centre, Mississippi State University, Stoneville.
Tucker CS, van der Ploeg M (1999) Management of Off-flavors in Channel Catfish Ponds. Fact Sheet 192. Southern Regional Aquaculture Center, Stoneville, Mississippi.
Tung SC, Lin TF, Liu CL, Lai SD (2004) The effect of oxidants on 2-MIB concentration with the presence of cynobacteria. Water Sci Technol 49:281–288.
Tyler L, Acree T, Butts R (1978) Odour characterization of the synthetic stereoisomers of 2-methylisoborneol. J Agric Food Chem 26:1415–1417.
Utkilen H, Frøshaug M (1992) Geosmin production and excretion in a planktonic and benthic *Oscillatoria*. Water Sci Technol 25(2):199–206.
van Breeman LWCA, Ketelaars HAM, Visser P, Ebbeng JH (1991) A new method to control growth of geosmin-producing benthic Cyanobacteria. Verh Int Ver Limnol 24:2168–2173.
van Breeman LWCA, Dits JS, Ketelaars HAM (1992) Production and reduction of geosmin and 2-methylisoborneol during storage of river water in deep reservoirs. Water Sci Technol 25(2):233–240.
van der Ploeg M (1991) Testing flavor quality of pre-harvest channel catfish. Publication Fact Sheet 431, Southern Regional Aquaculture Center, Stoneville, MS.
van der Ploeg M, Tucker CS, Boyd CE (1992) Geosmin and 2-methylisoborneol production by cyanobacteria in fish ponds in the southeastern United States. Water Sci Technol 25(2):283–290.
van der Ploeg M, Dennis ME, de Regt MQ (1995) Biology of *Oscillatoria* cf. *chalybea*, a 2-methylisoborneol producing blue-green alga of Mississippi catfish ponds. Water Sci Technol 31(11):173–180.
Veijanen A, Paasivirta J, Lahtipera M (1988) Structure and sensory analyses of tainting substances in Finnish freshwater environments. Water Sci Technol 20:43.
Veijanen A, Kolemainen E, Kauppinen P, Lahtiperä M, Paasivirta J (1992) Methods for the identification of tainting terpenoids and other compounds from algae. Water Sci Technol 25:165–170.
Vilalta E, Guasch H, Muñoz I, Navarro E, Romaní A, Valero F, Rodriguez JJ, Alcaraz R, Sabater S (2003) Ecological factors that co-occur with geosmin production by benthic cyanobacteria. The case of the Llobregat River. Algol Stud 109:579–592.
Vilalta E, Guasch H, Muñoz I, Romaní A, Valero F, Rodriguez JJ, Alcaraz R, Sabater S (2004) Nuisance odours produced by benthic cyanobacteria in a Mediterranean river. Water Sci Technol 49:25–31.
Wagner KJ, Corbin WB, Hudak J (1999) Control of benthic algal mats in a water supply reservoir by alum treatment. In: Abstracts, North American Lake Management Society Symposium, Dec. 1–4, 1999, Reno, NV.
Walker HL, Higginbotham LR (2000) An aquatic bacterium that lyses cynobacteria associated with off-flavor of channel catfish (*Ictalurus punctatus*). Biol Control 18:71–78.
Wasowicz E, Kaminski E, Kollmannsberger H, Nitz S, Berger RG, Drawert F (1988) Volatile compounds of sound and musty wheat grains. Chem Microbiol Technol Lebensm 11:161–168.
Watson SB (1999) Outbreaks of taste/odour causing algal species: theoretical, mechanistic and applied approaches. PhD thesis. University of Calgary, Calgary, Alberta, Canada.
Watson SB (2003) Cyanobacterial and eukaryotic algal odour compounds: signals or by-products. A review of their biological activity. Phycologia 42:332–350.
Watson SB, Ridal J (2003) Periphyton: a primary source of widespread and severe taste and odour: Water Sci Technol 49:33–39.
Watson SB, Brownlee B, Satchwili T, Hargesheimer E (2000) Quantitative analysis of trace levels of geosmin and MIB in source and drinking water using headspace SPME. Water Res 34:2818–2828.
Westerhoff P, Rodriguez-Hernandez M, Baker L, Sommerfeld M (2005) Seasonal occurrence and degradation of 2-methylisoborneol in water supply reservoirs. Water Res 39(20):4899–4912.

Whitfield FB, Freeman DJ, Banister, PA (1981) Dimethyl trisulfide: an important off-flavor component in the Royal Red Prawn (*Hymnopteraeus sibogae*). Chem Ind 316–317.
Whitfield FB, Last JH, Shaw KJ, Tindale CR (1988) 2,6-Dibromophenol: the cause of an iodoform-like off flavor in some Austalian crustaceans. J Sci Food Agric 46:29–42.
Whitfield FB, Shaw KJ, Svoronos D (1994) Effect of the natural environment on the flavor of seafoods: the flavor of *Girella tricuspidata*. Dev Food Sci 35:417–420.
Wnorowski AU (1992) Tastes and odors in the aquatic environment: a review. Water S Afr 18:203–214.
Wnorowski AU, Scott WE (1992) Incidence of off-flavors in South African surface waters. Water Sci Technol 25:225–232.
Wu J, Jüttner F (1988) Differential partitioning of geosmin and 2-methylisoborneol between cellular constituents in *Oscillatoria tenuis*. Arch Microbiol 150:580–583.
Wu J, Ma P, Chou T (1991) Variation of geosmin content in *Anabaena* cells and its relation to nitrogen utilization. Arch Microbiol 157:66–69.
Yagi M (1988) Musty odour problems in Lake Biwa 1982–1987. Water Sci Technol 20(8/9):133–142.
Yagi O, Sugiura N, Sudo R (1981) Odorous compounds produced by *Streptomyces* in Lake Kasumigaura. Verh Int Ver Limnol 21:641–645.
Yagi M, Kajino M, Matsuo U, Ashitani K, Kita T, Nakamura T (1983) Odour problems in Lake Biwa. Water Sci Technol 15(6/7):311–321.
Yagi M, Nakashima S, Muramoto S (1988) Biological degradation of musty odor compounds, 2-methylisoborneol and geosmin, in a bioactivated carbon filter. Water Sci Technol 20:255–260.
Young CC, Suffet IH, Crozes G, Bruchet A (1999) Identification of a wood-hay odor-causing compound in a drinking water supply. Water Sci Technol 40:273–278.
Yurkowski M, Tabachek JL (1980) Geosmin and 2-methylisoborneol implicated as a cause of muddy odour and flavor in commercial fish from Cedar Lake, Manitoba. Can J Fish Aquat Sci 37:1449.
Zaitlin B, Watson SB (2006) Actinomycetes in relation to taste and odour in drinking water: Myths, tenets and truths. Water Res 40(9):1741–1753.
Zaitlin B, Watson S, Ridal J, Satchwill T, Parkinson D (2003) Actinomycetes in Lake Ontario: habitats and geosmin and MIB production. J Am Water Works Assoc 95(2):113–118.
Zimba PV, Tucker CS, Mischeke CC (2002) Short term effect of diuron on catfish pond ecology. North Am J Aquacult 64:16–23.
Zimmerman LR, Ziegler AC, Thurman EM (2002) Method of analysis and quality-assurance practices by US Geological Survey Organic Geochemistry Research Group. Determination of Geosmin and methylisoborneol in water using solid phase microextraction and gas chromatography/mass spectrometry. Open file report 02–337. U.S. Geological Survey, Boulder, Co. 12.

Mercury in the San Francisco Estuary

Christopher H. Conaway[1], Frank J. Black[1], Thomas M. Grieb[2], Sujoy Roy[2], and A. Russell Flegal[1]

I Introduction

A Purpose and Scope

Ever since the recognition of mercury as an environmental problem, San Francisco Estuary has been an active area of mercury research. It is little wonder that this is so: the estuary is in the middle of a region of mercury mineralization and historic mercury mining, and it is downstream of an area of historic gold mining where millions

[1]Department of Environmental Toxicology, University of California, Santa Cruz, CA 95064, U.S.A.

[2]Tetra Tech, Inc., Lafayette, CA 94549, U.S.A.

Communicated by Ronald S. Tjeerdema.

of kilograms of mercury were used. It is also a heavily urbanized area that once featured chloralkali facilities and numerous shipyards potentially contaminated with mercury-based paints. In addition, it is a drainage area for rich agricultural regions that may have seen substantial environmental applications of mercury insecticides and fungicides. In this review, we present a survey of literature on mercury contamination and biogeochemistry focusing on San Francisco Estuary. Our intent is to stimulate scientific questions addressing mercury contamination in this and other estuarine systems, as well as to describe the restoration and management efforts that accompany mercury-contaminated sites.

B Overview of the Problem

Before presenting work specific to mercury contamination in San Francisco Estuary, an overview of the environmental mercury problem and mercury chemistry is appropriate. There are many valuable reviews on this wider topic, with focuses on toxicological (Clarkson and Magos 2006), biogeochemical (Benoit et al. 2003; Fitzgerald and Lamborg 2003; Fitzgerald et al. 2007; Ravichandran 2004; Ullrich et al. 2001), ecological (Wiener et al. 2003), and microbiological aspects (Barkay et al. 2003).

Mercury is an environmental and human health concern largely because of the formation of methylmercury, particularly monomethylmercury (MMHg), which is bioaccumulated and biomagnified to toxic concentrations in higher trophic level organisms, including birds (Schwarzbach et al. 2006) and mammals (Wiener et al. 2003). It is a neurotoxin for humans, and effects have been noted in populations consuming fish (Clarkson and Magos 2006). In estuarine systems, sediments are a primary area of MMHg production (Mason et al. 2006). Sulfate-reducing bacteria are thought to be the principal methylators of mercury in anoxic estuarine sediment (Compeau and Bartha 1985), although iron-reducing bacteria have recently also begun receiving scrutiny (Kerin et al. 2006). The production of MMHg is, therefore, controlled by factors influencing the distribution of mercury between abiotic and biotic compartments, such as sulfur chemistry and organic matter, and by factors that control microbial activity, such as temperature and the availability of suitable organic matter for cellular respiration (Gilmour and Henry 1991; Hammerschmidt and Fitzgerald 2004; Heyes et al. 2006; King et al. 2001). Methylmercury produced in sediment that is exported to the water column can be bioaccumulated by phytoplankton or other organisms (Pickhardt and Fisher 2007) and biomagnified to higher trophic levels (Lawson and Mason 1998).

C Environmental Setting of San Francisco Estuary

An understanding of the setting of San Francisco Estuary is essential as a backdrop for this review. The monograph *San Francisco Bay: The Urbanized Estuary* is an older, but excellent description (Conomos 1979), as is the more recent *San Francisco*

Bay: The Ecosystem (Hollibaugh 1996). In addition, articles are available on the characteristics and circulation patterns in the estuary (Conomos et al. 1985), temporal fluctuation and time scales of variability of estuarine parameters (Cloern and Nichols 1985; Thomson-Becker and Luoma 1985), and anthropogenic modification of the estuary over time (Nichols et al. 1986). Some recent studies have covered water circulation, salinity, and nutrients (Kimmerer 2002; Monismith et al. 2002; Smith and Hollibaugh 2006); suspended sediment (Ganju et al. 2005; McKee et al. 2006; Ruhl et al. 2001; Schoellhamer 2002); organic carbon (Lesen 2006; Murrell and Hollibaugh 2000; Stepanauskas et al. 2005); marsh formation (Watson 2004); and sedimentation (Foxgrover et al. 2004; Jaffe and Foxgrover 2006; Jaffe et al. 1998).

San Francisco Estuary is a truly unique setting (Fig. 1). It is a natural, semienclosed body of water created by right-lateral movement on the San Andreas fault system (Hedgpeth 1979). It is the largest estuary on the California coast and is heavily urbanized (Nichols et al. 1986). Its circulation is controlled by tidal currents and freshwater flow, which is dominated by the distinctly Mediterranean climate in the region—dry summers and wet winters (Kimmerer 2002). San Francisco Estuary can be divided into two geochemically distinct subestuaries, the northern and southern reaches, which join in the Central Bay and connect to the Pacific Ocean via the Golden Gate (Flegal et al. 1991). The system has further been divided into six hydrographically distinct regions: Tributaries, Southern Sloughs, South Bay, Central Bay, Northern Estuary, including San Pablo Bay and Suisun Bay, and River-Delta (Conaway et al. 2007). Ninety percent of the annual freshwater inflow to the estuary enters via the northern reach through the delta formed by the convergence of the Sacramento-San Joaquin drainage basins, which includes most of the Coast Ranges, the Central Valley of California, and the western Sierra Nevada (Conomos et al. 1985). The Napa and Petaluma Rivers, which also drain to the northern reach, provide local drainage from the Coast Ranges, but their discharges are relatively small in comparison. In contrast, the southern reach receives only a small amount of freshwater input (<10% of the total freshwater input to the estuary), mostly from the Guadalupe River, Coyote Creek, and other small tributaries that locally drain the Coast Ranges and the Santa Clara Valley. Onto this physically and chemically complex system is superimposed an ecologically and biogeochemically complex mercury contamination issue, which has been the focus of many studies reviewed here.

II Issues Related to Mercury Contamination in San Francisco Estuary

Concerns about mercury in San Francisco Estuary center on human health and ecological effects on birds. The San Francisco Bay Regional Water Quality Control Board (SFRWQCB), which is tasked with the preservation of beneficial uses of the estuary, has determined that the estuary is impaired for mercury, in part because of the reported concentrations of mercury in fish tissue and bird eggs (SFRWQCB 2006). Studies on fish and ecotoxicological effects on birds both support this regulatory statement and highlight concerns of mercury toxicity.

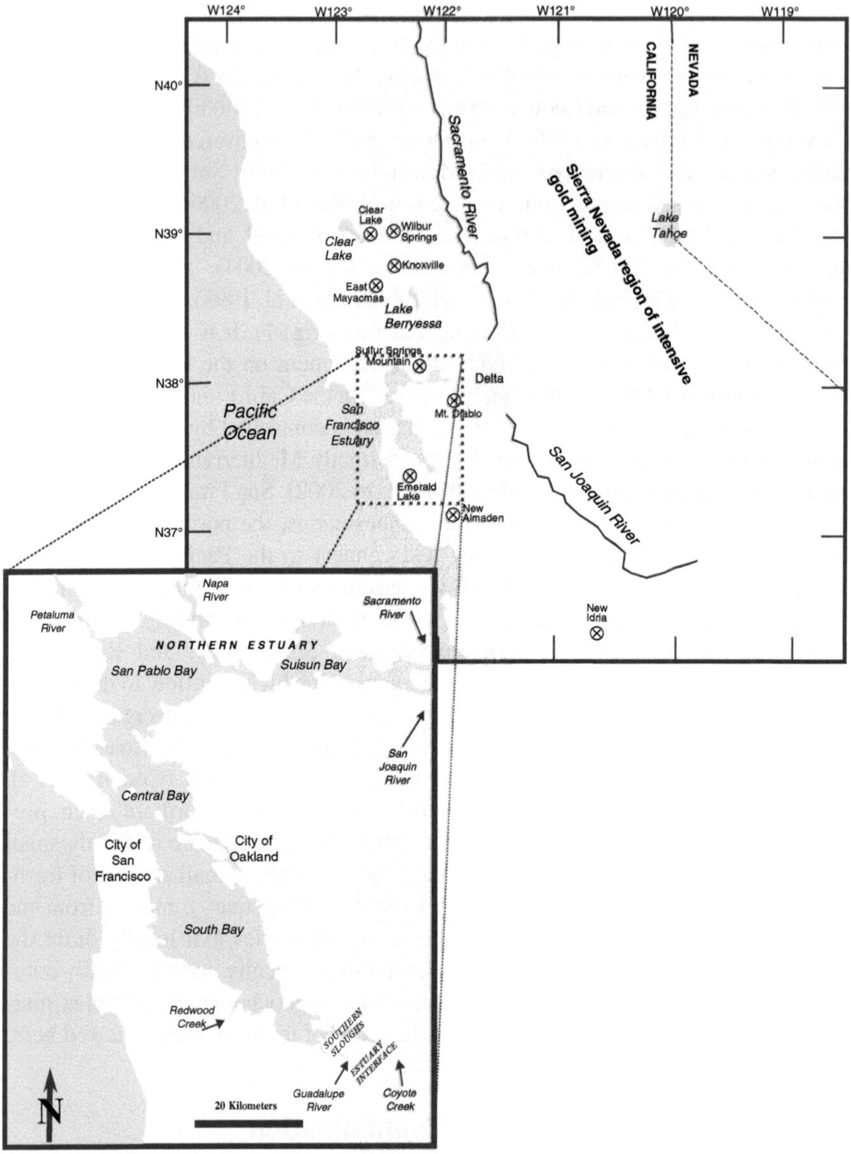

Fig. 1 Regional map of San Francisco Estuary, California, with inset detail. Regional map shows area where intensive gold mining in the foothills of the Sierra Nevada occurred. Locations of large mercury mining districts in the San Francisco Estuary watershed are shown with an "⊗" symbol. Inset shows the estuary and its larger tributaries. Distinct hydrographic regions are Rivers (the confluence of the Sacramento and San Joaquin), Northern Estuary, Central Bay, South Bay, the Southern Sloughs, and Estuary Interface

A Human Health

Consumption of mercury-contaminated fish from the estuary is the issue most relevant to human health. Accordingly, recent studies of mercury in fish in San Francisco Estuary (Davis et al. 2002; Fairey et al. 1997; Greenfield et al. 2005) have focused on concentrations and spatial and temporal trends in those concentrations in various fish species. The range of concentrations in several species are summarized in Table 1. Fish mercury concentrations can exceed regulatory standards in leopard shark, striped

Table 1 Survey of total mercury (Hg_T) and methylmercury (MeHg) concentrations ($\mu g\ g^{-1}$) in organisms from San Francisco Estuary

Species	Tissue	Hg_T ($\mu g\ g^{-1}$), range (mean), wet weight	MeHg, ($\mu g\ g^{-1}$), wet weight	Notes
Fish[a,b]				
California halibut (*Paralichthys californicus*)	Muscle	0.20–0.36		
Jacksmelt (*Atherinopsis californiensis*)	Muscle	0.068–0.17 (0.09)		
Leopard shark (*Triiakis semifasciata*)	Muscle	0.28–1.3		
Shiner surfperch (*Cymmatogaster aggregata*)	Muscle	0.068–0.42		
Striped bass (*Morone saxatilis*)	Muscle	0.15–0.55		
Sturgeon (*Acipenser transmontanus*)	Muscle	0.25–0.30		
White croaker (*Genyonemus lineatus*)	Muscle	0.069–0.41		
Birds				
California clapper rails *Rallus longirostris obsoletus*)[c]	Egg[d]	0.11–2.5		MeHg averaged 95% of total in subset analyzed
Canvasbacks (*Aythya valisineria*)[e]	Liver	ND–9.4[f]		
Greater scaup (*Aythya marila*)	Liver	1.8–20[f]		
Lesser scaup (*Aythya affinis*)	Liver	1.1–9.9[f]		
Surf scoters (*Melanitta perspicillata*)	Liver	5–21[f]		
Ruddy ducks (*Oxyura jamaicensis*)	Liver	2–7[f]		

(continued)

Table 1 (continued)

Species	Tissue	Hg_T (μg g^{-1}), range (mean), wet weight	MeHg, (μg g^{-1}), wet weight	Notes
Mammals				
Pacific harbor seal (*Phoca vitulina richardii*)[g]	Blood Hair Liver	0.015–1.4 0.41–93 0.15–160	0.068–2.9	Pups, juveniles, and adults from Central and Northern California, 2003–2005. Concentrations are typically highest in adults.
House mouse (*Mus musculus*)[h]	Liver	0.02–4.0		Small mammals collected from tidal salt marsh habitat dominated by pickleweed (*Salicornia virginica*).
Deer mouse (*Peromyscus maniculatus*)	Liver	0.05–1.1		
California vole (*Microtus caliJbrnicus*)	Liver	0.02–0.12		

ND, not detected.
[a] Guideline for human consumption is 0.23 μg g^{-1} (Davis et al., 2002).
[b] Davis et al. 2002; Fairey et al. 1997.
[c] Lonzarich et al. 1992; Schwarzbach et al. 2006.
[d] Lowest observed adverse effect concentrations (LOAEC) is 0.50 μg g^{-1} (Schwarzbach et al. 2006).
[e] Hoffman et al. 1998; Hothem et al. 1998; Takekawa et al. 2002.
[f] Dry weight.
[g] Brookens et al. 2007.
[h] Clark et al. 1992.

bass, and white sturgeon (Davis et al. 2002; Fairey et al. 1997; Greenfield et al. 2005). Concentrations in several species of fish are statistically higher in Oakland Harbor than in other locations in the estuary (Davis et al. 2002; Meador et al. 2005). Despite its drastic reduction in use over the past few decades, mercury concentrations in striped bass in the estuary show no long-term decrease since the 1970s (Greenfield et al. 2005).

To complement these studies on fish mercury concentration, a report on fish consumption in the San Francisco Bay area was conducted by the California Department of Health Services (SFEI 2000). Based on a survey of 1300 San Francisco Bay anglers representing 150 fishing sites visits, the study found that the five most popular fish eaten by anglers, in order of preference, were striped bass, California halibut, jacksmelt, white sturgeon, and white croaker, and that about 1 in 10 of the anglers ate more than the amount recommended by the California Environmental Agency's Office of Environmental Health Hazard Assessment (~6.2 g sportfish/kg body weight/mon). Similarly, a fish consumption study on low-income persons in the Sacramento-San Joaquin Delta region found that 2% of women exceeded the Office of Environmental Health Hazard Assessment recommendation, and that 29% of women exceeded the FDA/EPA advisory limits (48.6 g/d commercial fish, 24.3 g/d sportfish) via a combination of sport and commercial fish consumption (Silver et al. 2007).

In addition to studies related to fish consumption, there has been one published study (Windham et al. 2006) suggesting a potential association between children with autism spectrum disorders and emission of hazardous air pollutants, including mercury, to ambient air in the San Francisco Bay area. However, this association, based on limited data, should be reexamined and confirmed with a more refined exposure assessment.

B Ecosystem Health

Birds

Studies of mercury in birds in the estuary have focused on measuring concentrations and detecting the ecological impacts of contamination (Hoffman et al. 1998; Hothem et al. 1995, 1998; Hui 1998; Hui et al. 2001; Lonzarich et al. 1992; Ohlendorf et al. 1991; Schwarzbach et al. 2006; Takekawa et al. 2002); these have demonstrated potentially toxic mercury concentrations in waterbirds around the estuary (see Table 1), with diving ducks typically showing the highest level of contamination. The relationship between mercury and selenium has also been investigated because the elements may offset each other's toxicity (Hothem et al. 1998; Hui et al. 2001). Based on elevated concentrations of mercury, other recent studies of birds in the estuary have sought to establish biomarkers of exposure, such as enzyme activity and reproductive success. Ratios of glutathione reductase to oxidized glutathione were used as a bioindicator to discriminate between mercury and selenium effects (Hoffman et al. 1998). Schwarzbach et al. (2006) linked decreased egg hatchability to mercury contamination in the California clapper rail (*Rallus longirostris obsoletus*).

Mammals

There have been a few published studies on mercury concentration and toxicity in mammals. A study of small mammals residing in pickleweed (*Salicornia virginica*) habitats around the estuary showed relatively low mercury concentrations (<1 µg g^{-1} dry weight tissue) in salt marsh harvest mice (*Reithrodontomys raviventris*), house mice (*Mus musculus*), and deer mice (*Peromyscus maniculatus*), although concentrations did reach 4 µg g^{-1} in house mice at one site; these burdens had no observable effect on health of the mice (Clark et al. 1992). An investigation of mercury and MMHg in blood, hair, and liver of harbor seals (*Phoca vitulina richardii*) from central and northern California, including sites in the San Francisco Estuary area, showed concentrations that were considered toxic in other mammals (Brookens et al. 2007). However, the study did not have sufficient statistical power to resolve effects of sampling location on mercury concentrations (i.e., if seals from the estuary had higher concentrations than seals in adjacent coastal areas).

III Sources and History

The elevated concentrations of mercury in fish and birds in San Francisco Estuary are attributed to the widespread and persistent mercury contamination in the region (Greenfield et al. 2005; Schwarzbach et al. 2006). This contamination comes from both natural sources and anthropogenic activities, with origins ranging from local, to regional, to global scales. In this section, we review these sources and the transport of mercury to the estuary through water, sediment, and the atmosphere.

A Sources and Reservoir Size

Mineral Resources and Extraction

The estuary lies in the heart of the highly mineralized circum-Pacific mercury belt. The regional geology has been described in several sources, most notably the work of Bailey and Everhart (1964) and, more recently, Rytuba (2003). Within the California Coast Range part of this belt, major deposits are typically associated with serpentinites emplaced along fault zones and altered to an assemblage of silicate and carbonate minerals by carbon dioxide-rich hydrothermal fluids (Rytuba 2003). Background concentrations in the region are in the range of 10–100 µg g^{-1} (Conaway et al. 2004; Domagalski 2001; Hornberger et al. 1999; Kerin 2002; Thomas et al. 2002). Emissions of mercury vapor from natural background and mineralized areas in the region may contribute a sizeable amount of mercury to the atmosphere (Coolbaugh et al. 2002; Engle et al. 2006).

There are several large economic mercury deposits in the region, with locations and mining activities described in detail by Holmes (1965) and summarized by Cargill et al. (1980) and Rytuba (2000). The New Almaden mercury mining district, North America's largest, is situated 30 km south of the estuary and drains through small tributaries into South San Francisco Bay. Other large mercury mining districts—Clear Lake, Knoxville, East Mayacmas, and Wilbur Springs—range up to 100 km north of the estuary and drain eventually to the northern reach (now via the Yolo Bypass). The New Idria district, North America's second largest mercury producer, is more than 100 km south of the estuary in the San Joaquin River watershed. Post-1945 production of mercury at many of these locations was by reworking surface tailings (Holmes 1965). The legacy of these mining activities has been presented in studies addressing mercury speciation, chemical weathering, and erosion (Conaway et al. 2004; Domagalski et al. 2004; Ganguli et al. 2000; Kim et al. 2004; Lowry et al. 2004; Rytuba 2000, 2003; Slowey et al. 2005a; Thomas et al. 2002). It is unclear if contamination from any but the largest of these districts, New Almaden, has a great influence on mercury concentrations in the estuary itself; and despite the size and proximity of New Almaden to the estuary, there is little to suggest that New Almaden-derived contamination is a pervasive and overwhelming mercury source in the estuary compared to industrial activities and gold mining.

Use of Mercury in the Region

The majority of mercury produced in California in the late 19th and early 20th centuries was used in gold mining (James 2005; Nriagu 1994), principally in hydraulic mining and dredging activities in the California Sierra Nevada (Nriagu and Wong 1997). Contamination from this mining activity has occurred principally by hydraulic mining debris transported through the watershed to the estuary (Conaway et al. 2003; Hornberger et al. 1999; Hunerlach et al. 1999; Jaffe et al. 1998; Marvin-DiPasquale and Agee 2003). An estimated 12 million kg mercury was used for gold recovery in California, and 4.5 million kg was lost to the environment in placer mining operations throughout California (Alpers et al. 2005; Churchill 2000). Although the bulk of the hydraulic mining sediment reached the estuary near the turn of the 20th century (Hornberger et al. 1999), studies on upstream geomorphology and geochemistry of hydraulic mining sediment show that this is still a pervasive and actively eroding source of contamination (Hunerlach et al. 1999; James 2005; Savage et al. 2000; Slowey et al. 2005b).

By the mid-20th century, the use of mercury in gold recovery fell drastically, and the major use of mercury became the incorporation into electrical devices and at chloralkali facilities (Nriagu 1987). Environmental uses, such as antifouling paint, pesticides, fungicides, and slimicides for wastewater treatment, also represent the use of tens of thousands of kilograms of mercury per year in the United States: the authors are unaware of specific data for California. Between 1945 and 1970, more than 100,000 kg/yr of mercury was used in agricultural applications in the U.S. (Nriagu 1987), mainly in seed treatment and foliar applications (D'Itri 1972); however, the State of California did not require reporting of pesticide use by type until 1970 (Federighi 2001), making estimates of mercury use in agriculture difficult at best. Other industrial uses and sources are presented in Table 2.

Table 2 Some potential local and regional sources of mercury contamination to the San Francisco Estuary from historical and modern human activity

Industry	Notes
Chloralkali[a]	Chlorine and caustic soda produced by the Castner-Kellner process (mercury cell) at facilities in Oakland, CA (1919–1957) and Pittsburgh, CA (1917–1973). Contamination occurred potentially via wastewater, spills, and air.
Petrochemical[b]	Major center of refineries 1900–present. Mean concentration of mercury in CA crude oil is 0.011 $\mu g\ g^{-1}$, but some reported values exceed 1 $\mu g\ g^{-1}$. Although historically CA was self-reli ant in petroleum supplies, recent years (1996–present) show an increase in foreign imports. Fate of mercury in refining process not well known.
Medical waste incineration[c]	Facility operated in East Oakland 1982–2001. Large source of mercury to atmosphere during that time (800 kg yr^{-1} in 1996).
Wastewater[d]	Mercury used as slimicide in wastewater treatment. Wastewater also contains mercury lost from hospitals, dental offices, and industrial waste. Total amount unknown.
Gold mining[e]	Gold mining began in mid-19th century and continued into the 20th century; 12 M kg mercury used in hydraulic-placer mining, ore-processing at hard rock mines, and drift mines and dredging operations in CA, with 4.5 M kg mercury lost to the environment from hydraulic placer mines, and 1.4 M kg lost at hard rock mines.
Other sources[f]	Mercury released to the atmosphere from cement manufacturing plants, carbon black production facilities, and waste burning. Mercury disposal in landfills of electrical devices (the principal use of mercury in the 20th century), including batteries, switches, and lamps, likely represents a primary repository of mercury in the region.
Antifouling paint[g]	Use of mercury in antifouling paints in the estuary dates at least to early 20th century. Between 1940 and 1970, approximately 0.04 M kg mercury per year used in antifouling paint in the U.S. California-specific data not available.
Agriculture[h]	Seed and foliar application of mercury as pesticides and fungicides. Between 1945 and 1970, 100,000 kg per year of mercury was used in agricultural applications in U.S. Potential contamination from both Central Valley and local agriculture.
Mercury mining districts[i]	Production data presented to show relative activity in districts. Loss to the environment unknown. Potential contamination occurs primarily through the weathering and erosion of tailings and other waste rock.

Mining district name	Mercury produced (M kg)	Local watersheds upstream of estuary
Clear Lake (1870–1957)	4.4	Cache Creek
Wilbur Springs (1862–1961?)	1.9	Cache Creek
Knoxville (1862–1961?)	5.7	Cache Creek
East Mayacmas (1870–1956)	13	James and Putah Creeks, Lake Berryessa
Sulfur Springs (Vallejo) (1852–1943)	0.59	Blue Rock Springs Creek, Rindler Creek
Mt. Diablo (1863–1958)	0.38	Marsh Creek, San Joaquin River

(continued)

Table 2 (continued)

Mining district name	Mercury produced (M kg)	Local watersheds upstream of estuary
Emerald Lake (1955–1958)	0.09	Redwood Creek
New Almaden (1845–1975)	40	Guadalupe River
New Idria (1858–1972)	17	San Carlos Creek, San Joaquin River

[a] USEPA 2007b.
[b] Sheridan 2006; Wilhelm et al. 2007.
[c] CARB 2000.
[d] Nriagu 1987.
[e] Alpers et al. 2005; Churchill 2000; James 2005.
[f] CARB 2000; DTSC 2002; Nriagu 1987.
[g] Nriagu 1987; WHOI 1952.
[h] D'Itri 1972; Nriagu 1987.
[i] Cargill et al. 1980; Domagalski et al. 2004; Holmes 1965.

B Forces That Bring Mercury to the Estuary

Mercury contamination from historic mining and industrial sources reaches San Francisco Estuary via tributaries, wastewater input, and atmospheric deposition. The magnitude of flux from these various pathways was reviewed by MacLeod et al. (2005) and is presented in Table 3. Transport of suspended sediment by the Sacramento-San Joaquin Rivers into the northern reach is the primary input of mercury to the system (Domagalski 1998, 2001; Roth et al. 2001). This suspended sediment is dispersed throughout the estuary (Ruhl et al. 2001), but smaller tributaries, such as those draining the New Almaden Mining District, are important as well (Thomas et al. 2002). Preliminary data on the mercury isotopic composition of sediments, however, are insufficient to resolve different sources (Foucher and Hintelmann 2006). Mass balance calculations indicate that inputs of mercury from wastewater discharge are currently relatively small (Ellgas 2001; Hsu and Sedlak 2003), but may have been higher in the past before the advent of modern mercury removal technology from waste streams or when mercury compounds were used as slimicides in wastewater treatment (Nriagu 1987). Atmospheric deposition occurs through wet and dry deposition as a minor part of total flux to the estuary (Steding and Flegal 2002; Tsai and Hoenicke 2001). Mercury in precipitation in the region is higher than in adjacent coastal areas, which may represent scavenging of labile mercury from the atmosphere from local sources (Conaway et al. 2005; Steding and Flegal 2002). The relative bioavailability of the mercury from all these previously mentioned sources remains an important unknown.

IV Distribution, Speciation, and Transformation

The legacy of more than 150 years of mercury contamination has been distributed throughout the San Francisco Estuary, with current estuary surface sediment mercury concentrations roughly 5–15 times greater than background levels

Table 3 Estimate of recent or present-day fluxes of mercury (kg yr^{-1}) in San Francisco Estuary

	Mercury (kg yr^{-1})	Reference
Sources		
Atmospheric deposition		
Direct wet deposition	4.4–4.8	Steding and Flegal 2002; Tsai and Hoenicke 2001
Direct dry deposition	22.0	
Atmospheric emissions		
Stationary sources	244	CARB 2005
Areawide sources	1074	
On-road mobile	29	
Other mobile	83	
Gasoline combustion	0.7–13	Conaway et al. 2005
Watershed		
Central Valley	440–800	Domagalski 2001; McKee et al. 2005; SFRWQCB 2006; Thomas et al. 2002
Guadalupe River	4–116	
Wastewater	12	Ellgas 2001
Erosion of contaminated benthic sediments	460	SFRWQCB 2006
Stormwater runoff		
Urban	160	SFRWQCB 2006
Nonurban	25	
Sinks		
Ocean export	513	MacLeod et al. 2005
Burial	732	

(Conaway et al. 2004; Hornberger et al. 1999). The physical distribution and chemical speciation of this mercury in part determine the relationship between the sources and present-day human health and ecological effects described above. The distribution of MMHg and total mercury has been generally described by Choe and associates (Choe and Gill 2003; Choe et al. 2003), Heim et al. (2007), and Conaway et al. (2003). Focusing on the northern reach, the studies by Choe are distinguished by their detailed treatment of the surface water, including colloidal fractions, and are complemented by the study of Heim et al. which provides data on mercury speciation in sediment over an annual period. The study by Conaway et al. details total mercury and MMHg distribution in both water and sediment with a multiannual, multiseasonal statistical approach.

A general trend discernible from the data available is that water column concentrations of total mercury are higher in the rivers draining into the estuary than in the estuary itself. The northern reach, with large riverine inputs, has higher dissolved and total mercury concentrations than the southern reach, where so-called freshwater inputs are dominated by wastewater discharges. Total mercury concentrations in surficial sediments, averaging approximately 1 nmol g^{-1} (Choe et al. 2004; Conaway et al. 2003), are similar throughout the estuary as a result of mixing and resuspension, but decrease moving east into the delta (Heim et al. 2007). Water and sediment MMHg levels vary substantially both spatially and temporally, with highest

Table 4 Concentrations in water (ng L^{-1}), sediment (ng g^{-1}), and atmosphere (ng m^{-3}) in six hydrographic regions of San Francisco Estuary

	Water					Sediment		Air
Region	UHg_T (ng L^{-1})	FHg_T (ng L^{-1})	UMMHg (ng L^{-1})	FMMHg (ng L^{-1})	DGM (ng L^{-1})	Hg_T (ng g^{-1})	MMHg (ng g^{-1})	Hg^0 (ng m^{-3})
Rivers/Delta	2–10	0.4–2	0.04–0.3	0.02–0.08	0.04–0.2	20–500	0.02–0.08	
Northern Estuary	0.4–90	0.1–30	0.1	0.008–0.4	0.2–0.5	30–600	0.06–0.4	
Central Bay	0.3–10	0.08–0.6	0.02–0.06	0.02–0.1	0.01–0.04	10–400	0.0–0.7	2
South Bay	0.4–40	0.1–10	0.02–0.2	0.01–0.08	0.02–0.1	100–800	0.08–2	2
Southern Sloughs	6–70	0.1–4	0.08–0.5	0.3	0.1–0.2	70–800	0.2–2	
Estuary Interface	2–70	0.1–30	0.1–0.4	0.2	0.5–2	100–1000	0.6–3	

Total mercury in unfiltered water (UHg_T), total mercury in filtered water (FHg_T), monomethylmercury in unfiltered water (UMMHg), monomethylmercury in filtered water (FMMHg), dissolved gaseous mercury (DGM), total mercury in sediment (Hg_T), monomethylmercury in sediment (MMHg), and mercury vapor in air (Hg^0).

Sources: Data from Choe and Gill (2003), Choe et al. (2003), Conaway et al. (2003), and Conaway (2005).

concentrations found in and near wetlands (Choe et al. 2004; Marvin-DiPasquale et al. 2003) and in the central delta (Heim et al. 2007). Table 4 lists the concentration of mercury species in various matrices around the estuary.

Both Hg(II) and MMHg are highly particle reactive, with partition coefficients (K_d) of $10^{4.0}$–$10^{6.5}$ commonly measured in the estuary (Choe and Gill 2003; Choe et al. 2003; Conaway et al. 2003). As a result, much of the advective transport of mercury into and within the estuary occurs via particulate phases (Choe and Gill 2003; Choe et al. 2003; Conaway et al. 2003; Domagalski 2001; Roth et al. 2001) as mercury-sulfide minerals, adsorbed to particles, or associated with organic matter (Roth et al. 2001; Slowey et al. 2005b).

A Sediment: The Importance of Sediment Processes

A key area in linking mercury sources to ecological effects and human health is its biogeochemistry in estuarine sediments. Using samples from San Francisco Estuary, Olson and Cooper (1974, 1976) were the first to demonstrate that estuarine sediments were an important site for the methylation of Hg(II). Studies of microbial mercury methylation and demethylation potential in sediments from various environments throughout the estuary and delta (Marvin-DiPasquale and Agee 2003; Marvin-DiPasquale et al. 2003; Mehrotra and Sedlak 2005; Topping et al. 2004) have subsequently illustrated the role of wetlands as hotspots of mercury methylation as well as the importance of Hg(II) speciation and bioavailability, microbial community, and respiration rate in controlling the rates of microbially mediated MMHg production and degradation in sediments of different environments in the estuary and delta.

Although there has been an apparent decrease in total concentrations in surface sediment through time in some parts of the estuary, there has been no observed corresponding decrease of mercury concentrations in sportfish (Conaway et al. 2007). This discrepancy is primarily attributed to MMHg being the dominant form in fish (Kuwabara et al. 2007), and that mercury methylation and uptake is controlled by a complex interaction of various biogeochemical factors. Nevertheless, because of active erosion and redistribution of sediment in the estuary (Cappiella et al. 1999; Foxgrover et al. 2004; Jaffe and Foxgrover 2006), buried sediments with relatively higher mercury concentrations (Conaway et al. 2004; Hornberger et al. 1999; Marvin-DiPasquale and Agee 2003) may be remobilized and increase levels of bioavailable mercury.

B Water: The Importance of Flux and Complexation

The majority of studies on water column mercury have so far focused on the distribution and transport of contaminant mercury to and within the estuary, making regional model and mass balance calculations possible (MacLeod et al. 2005). Water column measurements used to support flux calculations in the estuary have established the importance of investigating concentrations in suspended matter entering the estuary and also the dominance of sediment resuspension on water column mercury concentrations (Conaway et al. 2003; Domagalski 2001; McKee et al. 2006). Evasion of dissolved gas mercury from surface waters appears to be a small flux out of the estuary (MacLeod et al. 2005); however, high-temporal resolution studies with better spatial resolution are still required (Conaway 2005). In addition to flux calculations, studies on the relationship between salinity and freshwater flow on dissolved concentrations in the water column (Choe and Gill 2003; Choe et al. 2003; Conaway et al. 2003) are important in understanding the uptake of mercury by organisms, and water column measurements of mercury and MMHg have been used to show that sediments are an important source to overlying water (Choe et al. 2004; Topping et al. 2004).

Despite the many studies of the biogeochemical cycling of mercury in the estuary, there are few studies to date describing the element's complexation and speciation, which are relevant to its bioavailability. Detailed investigations of this type in the surface waters have been limited to studies of surface water in tributaries and wastewater effluent that showed the presence of strong, or nonlabile, Hg(II)-complexing ligands with conditional stability constants similar to those of reduced sulfur-containing ligands (Black et al. 2007; Hsu and Sedlak 2003). However, further studies of the complexation and chemical speciation of Hg(II) and MMHg in surface waters or pore waters of the estuary, as well as riverine inputs to the estuary, are needed.

V Bioaccumulation and Biomagnification

Although some organisms in San Francisco Estuary have relatively high and potentially toxic concentrations, their routes of exposure are not well known, because few investigations have focused on trophic transfer of mercury in the estuary (e.g., Pickhardt et al. 2006). It is assumed that the predominant form in biota is MMHg, which is readily bioaccumulated at the base of food chains and then biomagnified in higher trophic levels in the estuary, as elsewhere (Wiener et al. 2003).

A *Bioaccumulation*

Measurements of mercury uptake by phytoplankton demonstrate the importance of mercury speciation in the estuary. Luengen (2007) reported a biodepletion of dissolved MMHg, but not the inorganic form, during a phytoplankton bloom in the southern reach of the estuary. This apparently selective uptake or scavenging of MMHg is consistent with mesocosm and lake studies showing an inverse correlation between algal abundance and MMHg concentrations in zooplankton and fish (Chen and Folt 2005; Pickhardt et al. 2002). Subsequent bloom decay in the estuary also appears to accelerate the formation of MMHg in suboxic benthic sediments (Luengen 2007).

In a phytoplankton culture experiment using two waters from the freshwater delta region of the estuary containing different concentrations of dissolved organic carbon (DOC), Pickhardt and Fisher (2007) showed greater bioaccumulation of added MMHg, which tended to accumulate in cytoplasm, than of added Hg(II), which accumulated in cell walls and membranes. Additional work using live and heat-killed cells suggested active uptake of MMHg in some phytoplankton. Pickhardt and Fisher also noted that bioaccumulation of MMHg was greater in high-DOC water, citing the possibilities of relatively higher neutral MMHg species in the higher-DOC water, or that the higher-DOC water enhanced phytoplankton membrane permeability to MMHg.

In contrast to this active uptake of MMHg by phytoplankton, some invertebrates in the estuary have been found to show relatively low bioaccumulation. Gunther et al. (1999) measured median accumulation factors of unity (0.9–1.3) for mercury in three species of filter-feeding bivalves (*Mytilus californianus*, *Crassostrea gigas*, and *Corbicula fluminea*) transplanted from relatively pristine sites in central California to contaminated sites within the estuary. This lack of bioaccumulation indicates that much of the mercury at the lower end of food chains in the estuary is not bioavailable (e.g., in inorganic forms) or is relatively dilute because of a bloom dilution effect. Similarly, concentrations measured in the tissues of a resident bivalve, *Macoma petalum*, at a tidal mudflat location in South San Francisco Bay are comparable to local sediment concentrations (Moon et al. 2005), demonstrating a lack of bioaccumulation.

Studies on trophic transfer of mercury and the effects of food web characteristics on its accumulation in organisms from San Francisco Estuary are few. Pickhardt et al.

(2006) contrasted the accumulation of additions of inorganic Hg(II) and MMHg from the dissolved phase and from invertebrate food by mosquitofish (*Gambusia affinis*) and redear sunfish (*Lepomis microlophus*) using water types collected from the delta with different DOC concentrations. Bioaccumulation and retention of MMHg in fish was substantially higher than that of inorganic mercury and differed by both fish species and water type. Based on biokinetic modeling of the observed fish concentrations, Pickhardt et al. concluded that high assimilation efficiencies and slow loss of MMHg from dietary sources are the principal determinants of mercury burdens, but that further research is needed to address the effects of DOC on its bioavailability at lower levels of the food web in aquatic ecosystems.

B Untangling Biomagnification: Food Webs and Environmental Effects

There are marked spatial and temporal variations in mercury concentrations in higher trophic level organisms in San Francisco Estuary, which confounds efforts to resolve processes governing bioaccumulation and biomagnification (Brookens et al. 2007; Greenfield et al. 2005). Interannual variation in sportfish in the estuary has been explained by factors ranging from variation in fish movement patterns, diet, and populations sampled, or, alternatively, variation in freshwater discharge causing increases of bioavailable mercury in the estuary (Greenfield et al. 2005).

Models of food chains need to be delineated within different regions and seasons in the estuary, and these modeling efforts are being complicated by ongoing stressors to the estuary (e.g., invasive species, climate change, water diversions and regulation, and wetland management and restoration). Although algal blooms have now been shown to have a bloom dilution effect on MMHg concentrations at the base of planktonic food chains in the estuary (Luengen 2007), what effect the new phenomenon of increasing phytoplankton blooms in the estuary (Cloern et al. 2006) will have on MMHg cycling in the estuary is unknown. These and other unknowns need to be addressed by first making accurate measurements of concentrations and speciation at different trophic levels and in different food chains within the estuary.

VI Management and Restoration

A History of Mercury Management in the Estuary

The history of mercury management in San Francisco Estuary begins with the widespread recognition of the environmental problem in the 1970s (D'Itri 1972). Measures resulting from litigation in the late 19th and early 20th century, such as the 1884 Sawyer Injunction and the 1893 Caminetti Act, sought to regulate the transport of hydraulic mining debris and had an impact on the distribution of mercury-contaminated sediment (James 2005), but mercury itself was not specifically considered. In the early 1970s, concentrations in striped bass and catfish in the estuary were found to exceed the federal guidelines, and fish consumption advisories were issued by the State Department of Health (NRC 1978). Between 1970 and 1990, economic pressure and regulatory measures by the U.S. Environmental Protection Agency (EPA), such as the cancellation of product registration for mercury-containing pesticides, led to the rapid decline of its use in the region (Sznopek and Goonan 2000). These regulatory measures and a decline in economic viability resulted in the closure of most of the region's mercury mines in the early to mid-1970s (Cargill et al. 1980). The New Almaden mining district closed in 1975 and is listed on the EPA's Abandoned Mine Lands CERCLIS ("Superfund") inventory (USEPA 2007a).

B Recent Management and Restoration Efforts

The thread of recent management efforts in the estuary begins in 1994, when the State of California's Bay Protection and Toxic Cleanup Program measured mercury concentrations in fish that humans consume from San Francisco Estuary (Fairey et al. 1997). Based on the results, a health advisory for consuming fish from the estuary caused the San Francisco Bay Regional Water Quality Control Board (SFRWQCB) to formally list San Francisco Bay as impaired by mercury (Davis et al. 2002). Since 1999, the SFRWQCB has been developing a Total Maximum Daily Load (TMDL) to determine the load reductions necessary to attain the water quality standard for mercury (Johnson and Looker 2003; SFRWQCB 2006). The development of the TMDL is based on data from the Regional Monitoring Program (RMP) for Water Quality in San Francisco Estuary that identified the magnitude of mercury contamination as well as the temporal and spatial variability of concentrations in water, sediment, and biota (Conaway et al. 2003, 2007; Hoenicke et al. 2003; Thompson et al. 2000). The results of numerous other research projects from San Francisco Estuary, already described, and other mercury-contaminated locations have provided the current basic understanding of the processes affecting mercury biogeochemistry in the estuary.

Published studies on the effects of mitigation or remediation on environmental mercury are notably lacking for San Francisco Estuary. A laboratory study by Mehrotra and Sedlak (2005) used iron additions to decrease mercury methylation in wetland sediments from locations surrounding the estuary. In addition, phytoremediation of mercury-contaminated sediments using water hyacinth (*Eichhornia crassipes*) has been evaluated in the delta region of the estuary (Greenfield et al. 2007; Riddle et al. 2002). Nonetheless, there is a dearth of published papers on mercury experiments using constructed, managed, or remediated wetlands for the estuary. Compounding this lack of information, the scale of design and implementation of wetland restoration activity in San Francisco Estuary is changing from small restoration activities to large, landscape-scale projects, such as the 60 km^2 South Bay Salt Pond Restoration Project (Simenstad et al. 2006).

Although San Francisco Estuary is one of the most studied estuaries in the world (Flegal et al. 2005), large uncertainties still remain regarding the processes and factors controlling mercury methylation and bioaccumulation within the system. One initial step in addressing these uncertainties was the development of a conceptual model of mercury in San Francisco Estuary (Tetra Tech 2006) by the Clean Estuary Partnership, a consortium of stakeholders from industry and municipalities, environmental organizations, resource management agencies, and academic institutions. The conceptual model used existing data to develop an overview of mercury biogeochemistry and also identified the key management questions (Table 5) that must be answered to meet the needs of resource managers and other stakeholders.

Table 5 Management questions, data requirements, and technical approaches identified in conceptual model of mercury

Management question	Key data requirements	Technical approach
What is the relative bioavailability of mercury from different sources to San Francisco Bay?	Chemical and physical form of the mercury from natural and anthropogenic sources to prioritize remediation strategies	• Direct measurements of MeHg • Mesocosm experiments to quantify bioavailability under controlled conditions
At what locations are current methylation rates and methylmercury flux highest?	Characterization of existing methylmercury pool in sediments to determine whether hotspots are present	• Direct measurements of net methylation rates • Simultaneous measurement of factors affecting methylation rate: DO, TOC, SO_4, H_2S, chloride
Can existing wetlands be managed or new wetlands be designed to minimize net methylation rates, or limit exposure to methylmercury that is produced?	Quantification of the response in bioaccumulation to wetland characteristics that can be fully or partially controlled, e.g., nutrient salinity, depth, vegetation levels, type, hydroperiod	• Mesocosm experiments to quantify bioavailability under controlled conditions
Given various scenarios for management actions, when will we likely see improvements in sediment and tissue concentrations?	Characterization of local and bay-wide sediment mercury concentrations in response to localized interventions; quantification of the effects of sediment deposition and erosion on estimates of recovery	• Localized interventions to remove or cap high-mercury sediments • Measurement of responses to localized interventions and mass-loading reductions • Dated, deep-sediment cores to estimate effects of sediment erosion on recovery
How should we best monitor to detect changes in mercury concentrations in sediments and tissue?	Detection of statistically significant changes in reliable indicators	• Measurements of total and methylmercury concentrations in surficial sediments in shallow, depositional areas • Characterization of mercury concentrations in fish indicator species

Source: Tetra Tech (2006).

Some of the management questions identified by the conceptual model have highlighted additional field studies required to fill current gaps in our knowledge and understanding (Table 5). Although extensive monitoring programs in the estuary provide a synoptic view of conditions (Hoenicke et al. 2003; Thompson et al. 2000), detailed information on areas where methylation rates and MMHg flux are highest is not currently available. If these could be identified, then they could either be targeted for restoration directly or managed such that they do not expand. Additionally, this information is needed to guide and evaluate the relative effectiveness of alternative corrective actions. Some of this information is becoming available as several large multi-year field projects are nearing completion in the region (e.g., Marvin-DiPasquale et al. 2005; Schwarzbach et al. 2005; Yee et al. 2005).

The implementation plan for mercury management developed by the SFRWQCB (2006) must also address the anticipated changes in and around the estuary. For example, the planned conversion of salt ponds to wetlands surrounding the estuary will restore unique habitat for biota, particularly for wading birds, and nursery grounds for many species (Goals Project 1999). However, as wetlands are hotspots of mercury methylation, there is a trade-off between the beneficial functions of wetlands and the environmental risk of increased MMHg production and export to the estuary. It is estimated that the response time of concentrations to changes in mercury loading to the estuary is several decades (MacLeod et al. 2005). To determine whether restoration actions over this time frame are achieving any benefits, new and effective monitoring strategies are required to identify locations and processes where changes are anticipated to occur over much shorter time frames.

Summary

This review presents some of the published and other important literature on mercury contamination in San Francisco Estuary. Studies on human consumption of contaminated sportfish and on detecting ecological impacts of this contamination in wetland areas validate concerns regarding mercury's toxicity in this system. Mining, industrial, and environmental uses of mercury have occurred for more than a century, resulting in its large historic and continuing transport to the estuary. Consequently, there is a widespread distribution in the estuary, but more work is needed to show its relative chemical and biological availability from these sources. The uptake of mercury in the estuary has been shown in phytoplankton, but studies on biomagnification in local food webs have yet to draw a clear path to impairment in sportfish and waterbirds. In light of these concerns of impairment and the need for further information, large restoration activities planned for the estuary will require new technical approaches to solve important management questions, such as the location of key areas of methylmercury production.

Acknowledgments Elizabeth Kerin, Richard Looker, and Ben Greenfield graciously provided comments on the text.

References

Alpers CN, Hunerlach MP, May JT, Hothem RL (2005) Mercury Contamination from Historical Gold Mining in California. Fact Sheet 2005–3014, Version 1.1. United States Geological Survey, Sacramento, CA.

Bailey EH, Everhart DL (1964) Geology and quicksilver deposits of the New Almaden district, Santa Clara County, California. 360, U.S. Geological Survey, Menlo Park, CA.

Barkay T, Miller SM, Summers AO (2003) Bacterial mercury resistance from atoms to ecosystems. FEMS Microbiol Rev 27:355–384.

Benoit J, Gilmour CC, Heyes A, Mason RP, Miller CL (2003) Geochemical and biological controls over methylmercury production and degradation in aquatic ecosystems. In: Cai Y, Braids OC (eds), Biogeochemistry of Environmentally Important Trace Elements. ACS Symposium Series No. 835. American Chemical Society, Washington, DC, pp 262–297.

Black FJ, Bruland KW, Flegal AR (2007) Competing ligand exchange-solid phase extraction method for the determination of the complexation of dissolved inorganic mercury (II) in natural waters. Anal Chim Acta 598:318–333.

Brookens TJ, Harvey JT, O'Hara TM (2007) Trace element concentrations in the Pacific harbor seal (*Phoca vitulina richardii*) in central and northern California. Sci Total Environ 372:676–692.

Cappiella K, Malzone C, Smith RE, Jaffe B (1999) Sedimentation and Bathymetry Changes in Suisun Bay: 1867–1990. Report 99–563. United States Geological Survey, Menlo Park, CA.

CARB (2000) California Toxics Inventory (revised 8/28/00). http://arb.ca.gov/toxics/cti/cti.htm (May 3, 2004).

CARB (2005) California Toxics Inventory. http://arb.ca.gov/toxics/cti/cti.htm (November 15, 2005).

Cargill SM, Root DH, Bailey EH (1980) Resource estimation from historical data: mercury, a test case. Stat Earth Sci 12:489–522.

Chen CY, Folt CL (2005) High plankton densities reduce mercury biomagnification. Environ Sci Technol 39:115–121.

Choe KY, Gill GA (2003) Distribution of particulate, colloidal, and dissolved mercury in San Francisco Bay estuary. 2. Monomethyl mercury. Limnol Oceanogr 48:1547–1556.

Choe KY, Gill GA, Lehman R (2003) Distribution of particulate, colloidal, and dissolved mercury in San Francisco Bay estuary. 1. Total mercury. Limnol Oceanogr 48:1535–1546.

Choe KY, Gill GA, Lehman RD, Han S, Heim WA, Coale KH (2004) Sediment-water exchange of total mercury and monomethyl mercury in the San Francisco Bay-Delta. Limnol Oceanogr 49:1512–1527.

Churchill R (2000) Contributions of mercury to California's environment from mercury and gold mining activities: insights from the historical record. EPA Conference on Assessing and Managing Mercury from Historic and Current Mining Activities, San Francisco, CA, November 28–30.

Clark DR, Foerster KS, Marn CM, Hothem RL (1992) Uptake of environmental contaminants by small mammals in pickleweed habitats at San Francisco Bay, California. Arch Environ Contam Toxicol 22:389–396.

Clarkson TW, Magos L (2006) The toxicology of mercury and its chemical compounds. Crit Rev Toxicol 36:609–662.

Cloern JE, Nichols FH (1985) Time scales and mechanisms of estuarine variability, a synthesis from studies of San Francisco Bay. Hydrobiologia 129:229–237.

Cloern JE, Jassby AD, Schraga TS, Dallas KL (2006) What is causing the phytoplankton increase in San Francisco Bay? In: The Pulse of the Estuary: Monitoring and Managing Water Quality in the San Francisco Estuary. San Francisco Estuary Institute, Oakland, pp 62–70.

Compeau GC, Bartha R (1985) Sulfate-reducing bacteria: principal methylators of mercury in anoxic estuarine sediment. Appl Environ Microbiol 50:498–502.
Conaway CH (2005) Sources, Transport, and Fate of Mercury in San Francisco Estuary and Tributaries. Ph.D. dissertation. University of California at Santa Cruz, CA.
Conaway CH, Squire S, Mason RP, Flegal AR (2003) Mercury speciation in the San Francisco Bay estuary. Mar Chem 80:199–225.
Conaway CH, Watson EB, Flanders JR, Flegal AR (2004) Mercury deposition in a tidal marsh of south San Francisco Bay downstream of the historic New Almaden mining district, California. Mar Chem 90:175–184.
Conaway CH, Mason RP, Steding DJ, Flegal AR (2005) Estimate of mercury emission from gasoline and diesel fuel consumption, San Francisco Bay area, California. Atmos Environ 39:101–105.
Conaway CH, Ross JRM, Looker R, Mason RP, Flegal AR (2007) Decadal mercury trends in San Francisco Estuary sediments. Environmental Research 105:53–66.
Conomos TJ (ed) (1979). San Francisco Bay: The Urbanized Estuary: Investigations into the Natural History of San Francisco Bay and Delta with Reference to the Influence of Man. Pacific Division of the American Association of the Advancement of Science, San Francisco, CA.
Conomos TJ, Smith RE, Gartner JW (1985) Environmental setting of San Francisco Bay. Hydrobiologia 129:1–12.
Coolbaugh MF, Gustin MS, Rytuba JJ (2002) Annual emissions of mercury to the atmosphere from natural sources in Nevada and California. Environ Geol 42:338–349.
Davis JA, May MD, Greenfield BK, Fairey R, Roberts C, Ichikawa G, Stoelting MS, Becker JS, Tjeerdema RS (2002) Contaminant concentrations in sport fish from San Francisco Bay, 1997. Mar Pollut Bull 44:1117–1129.
D'Itri FM (1972) The Environmental Mercury Problem. CRC Press, Cleveland, OH.
Domagalski J (1998) Occurrence and transport of total mercury and methyl mercury in the Sacramento River basin, California. J Geochem Explor 64:277–291.
Domagalski J (2001) Mercury and methylmercury in water and sediment of the Sacramento River Basin, California. Appl Geochem 16:1677–1691.
Domagalski JL, Alpers CN, Slotton DG, Suchanek TH, Ayers SM (2004) Mercury and methylmercury concentrations and loads in the Cache Creek watershed, California. Sci Total Environ 327:215–237.
DTSC (2002) Mercury Report, California Environmental Protection Agency Department of Toxic Substances Control, Sacramento, CA.
Ellgas B (2001) Comparison of 1999–2000 POTW Hg data, methods & special studies, CWEA/BADA Mercury Workshop, San Jose, CA.
Engle MA, Gustin MS, Goff F, Counce DA, Janik CJ, Bergfeld D, Rytuba JJ (2006) Atmospheric mercury emissions from substrates and fumaroles associated with three hydrothermal systems in the western United States. J Geophys Res Atmos 111: D17304.
Fairey R, Taberski K, Lamerdin S, Johnson E, Clark RP, Downing JW, Newman J, Petreas M (1997) Organochlorines and other environmental contaminants in muscle tissues of sportfish collected from San Francisco Bay. Mar Pollut Bull 34:1058–1071.
Federighi V (2001) Regulating Pesticides: The California Story. A Guide to Pesticide Regulation in California. DPR Publication No. 203. California Department of Pesticide Regulation, Sacramento, CA.
Fitzgerald W, Lamborg CH (2003) Geochemistry of mercury in the environment. In: Lollar BS (ed) Environmental Geochemistry. Treatise on Geochemistry. Elsevier, Oxford, pp. 107–148.
Fitzgerald WF, Lamborg CH, Hammerschmidt CR (2007) Marine biogeochemical cycling of mercury. Chem Rev 107:641–662.
Flegal AR, Smith GJ, Gill GA, Sanudo-Wilhelmy S, Anderson LCD (1991) Dissolved trace element cycles in the San Francisco Bay estuary. Mar Chem 36:329–363.
Flegal AR, Conaway CH, Scelfo GM, Hibdon S, Sanudo-Wilhelmy S (2005) Factors influencing measurements of decadal variations in metal contamination in San Francisco Bay, California. Ecotoxicology 14:645–660.

Foucher D, Hintelmann H (2006) High-precision measurement of mercury isotope ratios in sediments using cold-vapor generation multi-collector inductively coupled plasma mass spectrometry. Anal Bioanal Chem 384:1470–1478.

Foxgrover AC, Higgins SA, Ingraca MK, Jaffe BE, Smith RE (2004) Deposition, Erosion, and Bathymetric Change in South San Francisco Bay 1858–1983. Report 2004–1192. United States Geological Survey, Menlo Park, CA.

Ganguli PM, Mason RP, Abu-Saba KE, Anderson RS, Flegal AR (2000) Mercury speciation in drainage from the New Idria mercury mine, California. Environ Sci Technol 34:4773–4779.

Ganju NK, Schoellhamer DH, Bergamaschi BA (2005) Suspended sediment fluxes in a tidal wetland: measurement, controlling factors, and error analysis. Estuaries 28:812–822.

Gilmour CC, Henry EA (1991) Mercury methylation in aquatic systems affected by acid deposition. Environ Pollut 71:131–169.

Goals Project (1999) Baylands Ecosystem Habitat Goals. A report of habitat recommendations prepared by the San Francisco Bay Area Wetlands Ecosystem Goals Project, U.S. Environmental Protection Agency, San Francisco, Calif./S.F. Bay Regional Water Quality Control Board, Oakland, CA.

Greenfield BK, Davis JA, Fairey R, Roberts C, Crane D, Ichikawa G (2005) Seasonal, interannual, and long-term variation in sport fish contamination, San Francisco Bay. Sci Total Environ 336:25–43.

Greenfield BK, Siemering GS, Andrews JC, Rajan M, Andrews SP Jr, Spencer DF (2007) Mechanical Shredding of Water Hyacinth (*Eichhornia crassipes*): Impacts to Water Quality in the Sacramento-San Joaquin River Delta, California. Estuaries Coasts, 30:627–640.

Gunther AJ, Davis JA, Hardin DD, Gold J, Bell D, Crick JR, Scelfo GM, Sericano J, Stephenson M (1999) Long-term bioaccumulation monitoring with transplanted bivalves in the San Francisco Estuary. Mar Pollut Bull 38:170–181.

Hammerschmidt CR, Fitzgerald WF (2004) Geochemical controls on the production and distribution of methylmercury in near-shore marine sediments. Environ Sci Technol 38:1487–1495.

Hedgpeth JW (1979) San Francisco Bay—The Unsuspected Estuary: A History of Researches. In: Conomos TJ (ed), San Francisco Bay: The Urbanized Estuary Investigations into the Natural History of San Francisco Bay and Delta With Reference to the Influence of Man. Pacific Division of the American Association of the Advancement of Science, San Francisco, CA, 9–29.

Heim WA, Coale KH, Stephenson M, Choe K-Y, Gill GA, Foe C (2007) Spatial and habitat-based variations in total and methyl mercury concentrations in surficial sediments in the San Francisco Bay-Delta. Environ Sci Technol 41:3501–3507.

Heyes A, Mason RP, Kim E-H, Sunderland E (2006) Mercury methylation in estuaries: insights from using measuring rates using stable mercury isotopes. Mar Chem 102:134–147.

Hoenicke R, Davis JA, Gunther A, Mumley TE, Abu-Saba K, Taberski K (2003) Effective application of monitoring information: the case of San Francisco Bay. Environ Monit Assess 81:15–25.

Hoffman DJ, Ohlendorf HM, Marn CM, Pendleton GW (1998) Association of mercury and selenium with altered glutathione metabolism and oxidative stress in diving ducks from the San Francisco Bay region, USA. Environ Toxicol Chem 17:167–172.

Hollibaugh JT (1996) San Francisco Bay: The Ecosystem. Further Investigations into the Natural History of San Francisco Bay and Delta with Reference to the Influence of Man. Pacific Division AAAS, San Francisco, CA.

Holmes GH (1965) Mercury in California. In: Mercury Potential of the United States. Mercury. U.S. Department of the Interior, Bureau of Mines, Washington, DC.

Hornberger MI, Luoma SN, van Geen A, Fuller C, Anima R (1999) Historical trends of metals in the sediments of San Francisco Bay, California. Mar Chem 64:39–55.

Hothem RL, Roster DL, King KA, Keldsen TJ, Marois KC, Wainwright SE (1995) Spatial and temporal trends of contaminants in eggs of wading birds from San Francisco Bay, California. Environ Toxicol Chem 14:1319–1331.

Hothem RL, Lonzarich DG, Takekawa JE, Ohlendorf HM (1998) Contaminants in wintering canvasbacks and scaups from San Francisco Bay, California. Environ Monit Assess 50:67–84.

Hsu H, Sedlak DL (2003) Strong Hg(II) complexation in municipal wastewater effluent and surface waters. Environ Sci Technol 37:2743–2749.

Hui CA (1998) Elemental contaminants in the livers and ingesta of four subpopulations of the American coot (*Fulica americana*): an herbivorous winter migrant in San Francisco Bay. Environ Pollut 101:321–329.

Hui CA, Takekawa JY, Warnock SE (2001) Contaminant profiles of two species of shorebirds foraging together at two neighboring sites in south San Francisco Bay, California. Environ Monit Assess 71:107–121.

Hunerlach MP, Rytuba JJ, Alpers CN (1999) Mercury Contamination from Hydraulic Placer-Gold Mining in the Dutch Flat Mining District, California. Report 99–4018B. U.S. Geological Survey, Menlo Park, CA.

Jaffe BE, Foxgrover AC (2006) Sediment Deposition and Erosion in South San Francisco Bay, California from 1956 to 2005. Report 2006–1287. U.S. Geological Survey, Menlo Park, CA.

Jaffe BE, Smith RE, Torresan LZ (1998) Sedimentation and bathymetric change in San Pablo Bay, 1856–1983. Report 98–759. U.S. Geological Survey, Menlo Park, CA.

James LA (2005) Sediment from hydraulic mining detained by Englebright and small dams in the Yuba basin. Geomorphology 71:202–226.

Johnson B, Looker R (2003) Mercury in San Francisco Bay Total Maximum Daily Load (TMDL). Progress Report. California Regional Water Quality Control Board, San Francisco Bay Region, Oakland, CA.

Kerin EJ (2002) Mercury Concentrations and speciation in coastal stream sediments in Santa Cruz and San Mateo Counties: comparison of baseline levels to contaminated levels in San Francisco Bay. Senior Thesis, University of California at Santa Cruz, CA.

Kerin EJ, Gilmour CC, Roden E, Suzuki MT, Coates JD, Mason RP (2006) Mercury methylation by dissimilatory iron-reducing bacteria. Appl Environ Microbiol 72:7919–7921.

Kim CS, Rytuba JJ, Brown GE (2004) Geological and anthropogenic factors influencing mercury speciation in mine wastes: an EXAFS spectroscopy study. Appl Geochem 19:379–393.

Kimmerer WJ (2002) Physical, biological, and management responses to variable freshwater flow into the San Francisco estuary [Review]. Estuaries 25:1275–1290.

King JK, Kostka JE, Frischer ME, Saunders FM, Jahnke RA (2001) A quantitative relationship that demonstrates mercury methylation rates in marine sediments are based on the community composition and activity of sulfate-reducing bacteria. Environ Sci Technol 35:2491–2496.

Kuwabara JS, Arai Y, Topping BR, Pickering IJ, George GN (2007) Mercury speciation in piscivorous fish from mining-impacted reservoirs. Environ Sci Technol 41:2745–2749.

Lawson NM, Mason RP (1998) Accumulation of mercury in estuarine food chains. Biogeochemistry 40:235–247.

Lesen AE (2006) Sediment organic matter composition and dynamics in San Francisco Bay, California, USA: seasonal variation and interactions between water column chlorophyll and the benthos. Estuar Coast Shelf Sci 66:501–512.

Lonzarich DG, Harvey TE, Takekawa JE (1992) Trace element and organochlorine concentrations in California clapper rail (*Rallus longirostris obsoletus*) eggs. Arch Environ Contam Toxicol 23:147–153.

Lowry GV, Shaw S, Kim CS, Rytuba JJ, Brown GE (2004) Macroscopic and microscopic observations of particle-facilitated mercury transport from new idria and sulphur bank mercury mine tailings. Environ Sci Technol 38:5101–5111.

Luengen AC (2007) Investigating the spring bloom in San Francisco Bay: links between water chemistry, metal cycling, mercury speciation, and phytoplankton community composition. Dissertation thesis. University of California at Santa Cruz, CA.

MacLeod M, McKone TE, Mackay D (2005) Mass balance for mercury in the San Francisco Bay Area. Environ Sci Technol 39:6721–6729.

Marvin-DiPasquale M, Agee JL (2003) Microbial mercury cycling in sediments of the San Francisco Bay-Delta. Estuaries 26:1517–1528.

Marvin-DiPasquale MC, Agee JL, Bouse RM, Jaffe BE (2003) Microbial cycling of mercury in contaminated pelagic and wetland sediments of San Pablo Bay, California. Environ Geol 43:260–267.

Marvin-DiPasquale M, Stewart AR, Fisher NS, Pickhardt PC, Mason RP, Heyes A, Winham-Meyer L (2005) Evaluation of Mercury Transformations and Trophic Transfer in the San Francisco Bay/Delta: Identifying Critical Processes for the Ecosystem Restoration Program. Annual Report of Progress for Project # ERP-02-P40. California Bay Delta Authority, Sacramento, CA.

Mason RP, Kim E-H, Cornwell J, Heyes D (2006) An examination of the factors influencing the flux of mercury, methylmercury and other constituents from estuarine sediment. Mar Chem 102:96–110.

McKee L, Leatherbarrow J, Oram J (2005) Concentrations and loads of mercury, PCBs, and OC pesticides in the lower Guadalupe River, San Jose, California: Water Years 2003 and 2004. SFEI Contribution 409. San Francisco Estuary Institute, Oakland, CA.

McKee L, Ganju N, Schoellhamer DH (2006) Estimates of suspended sediment entering San Francisco Bay from the Sacramento and San Joaquin Delta, San Francisco Bay, California. J Hydrol 323:335–352.

Meador JP, Ernest DW, Kagley AN (2005) A comparison of the non-essential elements cadmium, mercury, and lead found in fish and sediment from Alaska and California. Sci Total Environ 339:189–205.

Mehrotra AS, Sedlak DL (2005) Decrease in net mercury methylation rates following iron amendment to anoxic wetland sediment slurries. Environ Sci Technol 39:2564–2570.

Monismith SG, Kimmerer W, Burau JR, Stacey MT (2002) Structure and flow-induced variability of the subtidal salinity field in northern San Francisco Bay. J Phys Oceanogr 32:3003–3019.

Moon E, Shouse MK, Parchaso F, Thompson JK, Luoma SN, Cain DJ, Hornberger MI (2005) Near-field receiving water monitoring of trace metals and a benthic community near the Palo Alto Regional Water Quality Control Plant in south San Francisco Bay, California: 2004. OFR 2005–1279. United States Geological Survey, Menlo Park, CA.

Murrell MC, Hollibaugh JT (2000) Distribution and composition of dissolved and particulate organic carbon in northern San Francisco Bay during low flow conditions. Estuar Coast Shelf Sci 51:75–90.

Nichols FH, Cloern JE, Luoma SN, Peterson DH (1986) The modification of an estuary. Science 231:567–573.

NRC (1978) An Assessment of Mercury in the Environment. National Research Council, Washington, DC.

Nriagu JO (1987) Production and uses of mercury. In: Nriagu JO (ed) The Biogeochemistry of Mercury in the Environment. Topics in Environmental Health. Elsevier, New York, pp 23–40.

Nriagu JO (1994) Mercury pollution from the past mining of gold and silver in the Americas. Sci Total Environ 149:167–181.

Nriagu JO, Wong HKT (1997) Gold rushes and mercury pollution. In: Sigel A, Sigel H (eds) Mercury and Its Effects on Environment and Biology. Metal Ions in Biological Systems. Dekker, New York, pp 131–160.

Ohlendorf HM, Marois KC, Lowe RW, Harvey TE, Kelly PR (1991) Trace elements and organochlorines in surf scoters from San Francisco Bay, 1985. Environ Monit Assess 18:105–122.

Olson BH, Cooper RC (1974) In situ methylation of mercury in estuarine sediments. Nature (Lond) 252:682–683.

Olson BH, Cooper RC (1976) Comparison of aerobic and anaerobic methylation of mercuric chloride by San Francisco Bay sediments. Water Resour 10:113–116.

Pickhardt PC, Fisher NS (2007) Accumulation of inorganic and methylmercury by freshwater phytoplankton in two contrasting water bodies. Environ Sci Technol 41:125–131.

Pickhardt PC, Folt CL, Chen CY, Klaue B, Blum JD (2002) Algal blooms reduce the uptake of toxic methylmercury in freshwater food webs. Proc Natl Acad Sci USA 99:4419–4423.

Pickhardt PC, Stepanova M, Fisher NS (2006) Contrasting uptake routes and tissue distributions of inorganic and methylmercury in mosquitofish (*Gambusia affinis*) and redear sunfish (*Lepomis microlophus*). Environ Toxicol Chem 25:2132–2142.

Ravichandran M (2004) Interactions between mercury and dissolved organic matter: a review. Chemosphere 55:319–331.

Riddle SG, Tran HH, Dewitt JG, Andrews JC (2002) Field, laboratory, and X-ray absorption spectroscopic studies of mercury accumulation by water hyacinths. Environ Sci Technol 36: 1965–1970.

Roth DA, Taylor HE, Domagalski J, Dileanis P, Peart DB, Antweiler RC, Alpers CN (2001) Distribution of inorganic mercury in Sacramento River water and suspended colloidal sediment material. Arch Environ Contam Toxicol 40:161–172.

Ruhl CA, Schoellhamer DH, Stumpf RP, Lindsay CL (2001) Combined use of remote sensing and continuous monitoring to analyze variability of suspended-sediment concentrations in San Francisco Bay, California. Estuar Coast Shelf Sci 53:801–812.

Rytuba JJ (2000) Mercury mine drainage and processes that control its environmental impact. Sci Total Environ 260:57–71.

Rytuba JJ (2003) Mercury from mineral deposits and potential environmental impact. Environ Geol 43:326–338.

Savage KS, Bird DK, Ashley RP (2000) Legacy of the California Gold Rush: environmental geochemistry of arsenic in the southern Mother Lode Gold District. Int Geol Rev 42: 385–415.

Schoellhamer DH (2002) Variability of suspended-sediment concentration at tidal to annual time scales in San Francisco Bay, USA. Cont Shelf Res 22:1857–1866.

Schwarzbach SE, Suchanek TH, Heinz GH, Ackerman JT, Eagles-Smith CA, Adelsbach TL, Takekawa JY, Miles AK, Hoffman DJ, Cruz SEW-DL, Spring SE, Ricca MA, Maurer TC (2005) Mercury in birds of the San Francisco Bay-Delta: trophic pathways, bioaccumulation and ecotoxicological risk to avian reproduction. Unpublished 2005 annual report. U.S. Geological Survey, Western Ecological Research Center, and U.S. Fish and Wildlife Service, Environmental Contaminants Division, Sacramento, CA.

Schwarzbach SE, Albertson JD, Thomas CM (2006) Effects of predation, flooding, and contamination on reproductive success of California Clapper Rails (*Rallus longirostris obsoletus*) in San Francisco Bay. Auk 123:45–60.

SFEI (2000) San Francisco Bay Seafood Consumption Study. San Francisco Estuary Institute, Richmond, CA.

SFRWQCB (2006) Mercury in San Francisco Bay. Proposed Basin Plan Amendment and staff report for revised Total Maximum Daily Load (TMDL and proposed mercury water quality objectives. San Francisco Bay Regional Water Quality Control Board, Oakland, CA.

Sheridan M (2006) California Crude Oil Production and Imports. Staff Report CEC-600-2006-006. California Energy Commission Fossil Fuels Office, Sacramento, CA.

Silver E, Kaslow J, Lee D, Lee S, Lynn Tan M, Weis E, Ujihara A (2007) Fish consumption and advisory awareness among low-income women in California's Sacramento-San Joaquin Delta. Environ Res 104:410–419.

Simenstad C, Reed D, Ford M (2006) When is restoration not? Incorporating landscape-scale processes to restore self-sustaining ecosystems in coastal wetland restoration. Ecol Eng 26:27–39.

Slowey AJ, Johnson SB, Rytuba JJ, Brown GE (2005a) Role of organic acids in promoting colloidal transport of mercury from mine tailings. Environ Sci Technol 39:7869–7874.

Slowey AJ, Rytuba JJ, Brown GE (2005b) Speciation of mercury and mode of transport from placer gold mine tailings. Environ Sci Technol 39:1547–1554.

Smith SV, Hollibaugh JT (2006) Water, salt, and nutrient exchanges in San Francisco Bay. Limnol Oceanogr 51:504–517.

Steding DJ, Flegal AR (2002) Mercury concentrations in coastal California precipitation: evidence of local and trans-Pacific fluxes of mercury to North America. J Geophys Res Atmos 107:D24.

Stepanauskas R, Moran MA, Bergamasch BA, Hollibaugh JT (2005) Sources, bioavailability, and photoreactivity of dissolved organic carbon in the Sacramento-San Joaquin River Delta. Biogeochemistry 74:131–149.

Sznopek JL, Goonan TG (2000) The Materials Flow of Mercury in the Economies of the United States and the World. Report 1197. U.S. Geological Survey, Denver, CO.

Takekawa JY, Wainwright-De La Cruz SE, Hothem RL, Yee J (2002) Relating body condition to inorganic contaminant concentrations of diving ducks wintering in coastal California. Arch Environ Contam Toxicol 42:60–70.

Tetra Tech (2006) Conceptual Model for Mercury in San Francisco Bay. Prepared for the Clean Estuary Partnership by Tetra Tech, Inc., Lafeyette, CA.

Thomas MA, Conaway CH, Steding DJ, Marvin-DiPasquale M, Abu-Saba KE, Flegal AR (2002) Mercury contamination from historic mining in water and sediment, Guadalupe River and San Francisco Bay, California. Geochem Explor Environ Anal 2:211–217.

Thompson B, Hoenicke R, Davis JA, Gunther A (2000) An overview of contaminant-related issues identified by monitoring in San Francisco Bay. Environ Monit Assess 64:409–419.

Thomson-Becker EA, Luoma SN (1985) Temporal fluctuations in grain size, organic materials and iron concentrations in intertidal surface sediment of San Francisco Bay. Hydrobiologia 129:91–107.

Topping BR, Kuwabara JS, Marvin-DiPasquale MC, Agee JL, Kieu LH, Flanders JR, Parchaso F, Hager SW, Lopez CB, Krabbenhoft DP (2004) Sediment remobilization of mercury in South San Francisco Bay, California. Report 2004–5196. U.S. Geological Survey, Menlo Park, CA.

Tsai P, Hoenicke R (2001) San Francisco Bay Atmospheric Deposition Pilot Study Part 1: Mercury. San Francisco Estuary Institute Regional Monitoring Program for Trace Substances, Richmond, CA.

Ullrich SM, Tanton TW, Abdrashitova SA (2001) Mercury in the aquatic environment: a review of factors affecting methylation. Crit Rev Environ Sci and Technol 31:241–293.

USEPA (2007a) AML CERCLIS Inventory. http://www.epa.gov/superfund/programs/aml/aml-site/nonnpl.htm (June 2, 2007).

USEPA (2007b) Cleanup of Mercury-Contaminated Chlor-alkali Sites. http://www.epa.gov/epa-oswer/hazwaste/mercury/cleanup.htm (June 1, 2007).

Watson EB (2004) Changing elevation, accretion, and tidal marsh plant assemblages in a South San Francisco Bay tidal marsh. Estuaries 27:684–698.

WHOI (1952) Marine fouling and its prevention. Prepared by Woods Hole Oceanographic Institution for United States Naval Institute, Annapolis, MD.

Wiener JG, Krabbenhoft D, Heinz GH, Scheuhammer AM (2003) Ecotoxicology of mercury. In: Hoffman DJ, Rattner BA, Burton GAJ, Cairns JJ (eds) Handbook of Ecotoxicology. CRC Press, Boca Raton, FL, pp 409–463.

Wilhelm SM, Liang L, Cussen D, Kirchgessner DA (2007) Mercury in crude oil processed in the United States (2004). Environ Sci Technol 41:4509–4514.

Windham GC, Zhang LX, Gunier R, Croen LA, Grether JK (2006) Autism spectrum disorders in relation to distribution of hazardous air pollutants in the San Francisco Bay area. Environ Health Perspect 114:1438–1444.

Yee D, Collins J, Grenier J, Takekawa JE, Schwarzbach SE, Marvin-DiPasquale M, Krabbenhoft D, Evens J (2005) Mercury and Methylmercury Processes in North San Francisco Bay Tidal Wetland Ecosystems. Annual Report of Progress for Project #ERP-02D-P64. California Bay-Delta Authority, Sacramento, CA.

DDE Remediation and Degradation

John E. Thomas[1], Li-Tse Ou[1], and Abid Al-Agely[1]

I Introduction

DDE (2,2-bis(*p*-chlorophenyl)-1,1-dichloroethylene) is not a natural product; it is found only as a recalcitrant degradation product of 2,2-bis(*p*-chlorophenyl)-1,1,1-trichloroethane (DDT) or as a contaminant in technical-grade DDT (Metcalf 1995). DDT was produced and used starting in 1939, finally gaining widespread use by 1943 (Turusov et al. 2002). By the 1960s, evidence indicated that DDT and its metabolites, DDE and DDD (1,1-dichloro-2,2-bis(*p*-chlorophenyl)-ethane), were highly persistent in the environment and accumulated in higher mammals. DDT derivatives have been reported to be responsible for the thinning of bird eggshells (Heberer and Dunnbier 1999). Other studies suggested that the derivatives and isomers of DDT are endocrine disrupters causing impaired reproduction in wildlife by emasculation and abnormal sexual development (Sharpe 1995). The U.S. Environment Protection Agency (USEPA) has determined that DDT, DDD, and DDE are probable human carcinogens (ATSDR 2002). DDT was banned in the United States during the early 1970s, except for the emergency control of vector-borne diseases (Heberer and Dunnbier 1999; Spencer et al. 1996).

Technical-grade DDT, the most common formulation to be used as an insecticide, is composed of 14 chemicals with only 65%–80% being the active ingredient, *p,p*'-DDT. Technical DDT also contains 15%–21% of the inactive *o,p*'-DDT, 4% DDD, and up to 1.5% of 1-(*p*-chlorophenyl)-2,2,2-trichloroethanol. DDE is not an appreciable

[1]University of Florida, Soil and Water Science Deptartment, Gainesville, FL 32611, U.S.A.

ingredient in technical DDT but is a contaminant (Metcalf 1995). Although the use of DDT in the U.S. was discontinued in 1973, DDT and its metabolites are frequently encountered in soil, with *p,p*'-DDE being a major component (Aigner et al. 1998; ATSDR 2002; Boul et al. 1994; Thiele et al. 1997). Although DDE can come from technical-grade DDT as a contaminant, the production of DDE from DDT may result from aerobic biotic degradation, abiotic dehydrochlorination, and even from photochemical decomposition (Thiele et al. 1997). DDE, however, has been reported to be more persistent than DDT and can be found in soil decades after the last DDT treatment (Boul et al. 1994; Spencer at al. 1996). The U.S. Geological Survey (USGS 1999) has reported that in the U.S. the frequency of detection was *p,p*'-DDE (in 60% of urban sites and 48% of rural areas surveyed), followed by *p,p*'-DDD, *p,p*'-DDT, *o,p*'-DDD, *o,p*'-DDT, and *o,p*'-DDE, with the latter two being in less than 5% of the samples.

DDE toxicity and recalcitrance to degradation are regarded by many as a serious environmental problem. To be able to model and predict the fate and transport of DDE, it is necessary to have accurate partitioning data. However, the reported values of the physicochemical properties often vary by several orders of magnitude for *p,p*'-DDE; hence, the values reported in Table 1 represent the means (Shen and Wania 2005). Data for *o,p*'-DDE were scarce, with the most recent values given in ATSDR (2002).

The formula $\log K_{oc} = 0.989 \log (K_{ow}) - 0.346$ uses the octanol–water partition coefficient to give a semiempirical measurement of the sorption of hydrophobic chemicals to soil and sediment (Karickhoff 1981). For *p,p*'-DDE and *o,p*'-DDE, the log K_{oc} values are 3.1 and 2.6, respectively. A value greater than 3 indicates a strongly sorbed chemical with little potential for leaching from the soil surface. However, the chemical may be bound to soluble humic acid material that occurs as free-moving organic carbon and can

Table 1 Data available for *p,p*'-DDE (Shen and Wania 2005) and *o,p*'-DDE (ATSDR 2002)

Common name	*p,p*'-DDE	*o,p*'-DDE
IUPAC name	1-Chloro-4-[2,2-dichloro-1-(4-chlorophenyl)ethenyl] benzene	1-Chloro-2-[2,2-dichloro-1-(4-chlorophenyl)ethenyl] benzene
Structure	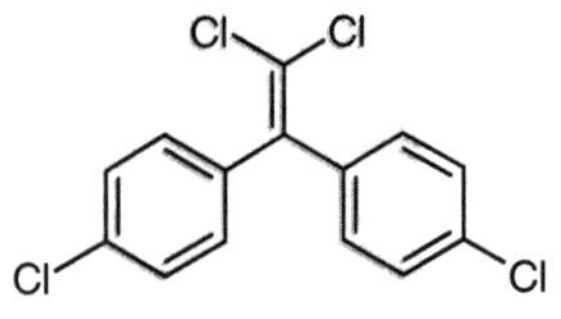	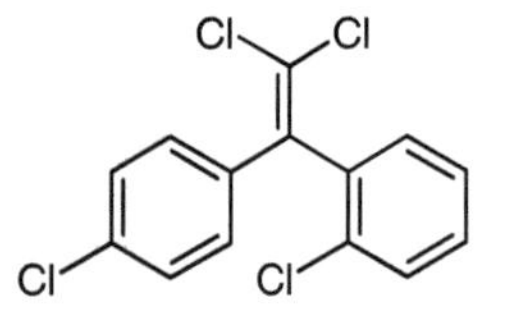
CAS registry number	72-55-9	3424-82-6
Molecular weight (g/mole)	318.03	318.03
Melting point	88.6°C	No data
Aqueous solubility (mg/L at 25°C)	0.26	0.14
Vapor pressure (torr at 25°C)	4.3×10^{-6}	6.2×10^{-6}
Henry's law constant (atm-m^3/mol at 25°C)	4.1×10^{-5}	1.8×10^{-5}
Log (octanol–water partition coefficient)	6.96	6.00

act as a carrier to facilitate the transport into subsurface soils despite the low aqueous solubility (Ding and Wu 1997). Dispersion of DDE may occur through its adsorption onto particulates, such as soil colloids, that are associated with the clay fraction that can be eroded and carried to streams as runoff (Masters and Inman 2000). Another loss mechanism is by volatilization from soil and water, with the trend being predicted by Henry's law constant. The half-lives of *p,p*'- and *o,p*'-DDE from a model river that is 1 m deep, flowing at 1 m/sec, with a wind of 3 m/sec, are 3.3 and 3.7 d, respectively (ATSDR 2002). It has also been reported, based upon laboratory study of the air–water partitioning, that DDE will vaporize 10–20 times faster from seawater than from freshwater (Atlas et al. 1982). Soil–air exchanges of DDE can also occur, with one model predicting that 200–600 kg *p,p*'-DDE is released from Alabama soil each year (Harner et al. 2001). It has also been reported that *p,p*'-DDE comprised 66% of the total DDT residuals in the atmosphere over a field that had been treated with DDT over a 7-yr period then untreated for the next 2 yr; this suggests that volatilization by degradation products can be a major pathway for loss by some organochorine insecticides in soil (Cliath and Spencer 1972; Hussain et al. 1994).

In contrast, the half-life for reduction in atmospheric concentrations has been measured in the Great Lakes area of the U.S. as ranging from 3.8 to 6.0 yr (Cortes and Hites 2000), and with no decrease measured for *p,p*'-DDE in the Canadian arctic (Hung et al. 2002), which indicates a temperature dependence. The half-life of *p,p*'-DDE in soil treated one time has been given as 5.7 yr (Beyer and Krynitsky 1989); however, in soil that had repeated DDT applications, the amount of extractable *p,p*'-DDE had not appreciably changed in 20 yr (Boul et al. 1994). It was presumed that any DDE losses from the latter soil were compensated by further transformation of *p,p*'-DDT. Therefore, to negate the toxicity and persistence of DDE in soil that has been repeatedly treated, some form of remediation must take place. Research into the remediation of DDE in soil and water has been done predominantly using (1) phytoremediation (phytoextraction), (2) aerobic biodegradation, (3) anaerobic biodegradation, and (4) abiotic degradation.

II Phytoremediation

Most of the literature on the phytoremediation of DDE focuses on the translocation of DDT or DDE from soil or water into plants. Plant species that have been investigated include rye, mustard, canola, vetch, pigeonpea, clover, peanut, white lupin, chicory, squash, cucumber, pumpkin, zucchini, tall fescue, leek, duckweed, parrot feather, and elodea (Gao et al. 2000; Gonzalez et al. 2003; Lunney et al. 2004; Suresh et al. 2005; White 2002; White et al. 2005). For rye, vetch, pigeonpea, clover, and white lupin, reductions or nonsignificant changes in *p,p*'-DDE uptake were observed when the nutrient treatment was varied or when there was a change in the plant biomass. In contrast, the amount of *p,p*'-DDE extracted from the soil doubled and was directly correlated to the plant biomass for mustard, canola, and peanut (White et al. 2005). The idea that fertilizer enhances phytoremediation appears to be highly species

Table 2 The five most effective plants assessed in phytoremediation studies and their associated reductions in DDE

Cultivar variety	Genus, species, subspecies	Percent uptake of DDE	Soil amendment	Reference
Connecticut Field (pumpkin)	*Cucurbita pepo* ssp. *pepo*	6.0	Mycorrhizal inoculant	White et al. 2006b
Howden (pumpkin)	*C. pepo* ssp. *pepo*	2.4	None	White 2002
Goldrush (zucchini)	*C. pepo* ssp. *pepo*	2.1	0.01 M citric acid	White et al. 2003b
Raven (zucchini)	*C. pepo* ssp. *pepo*	1.4	Mycorrhizal inoculant	White et al. 2006b
Costata Romanesco (zucchini)	*C. pepo* ssp. *pepo*	1.1	Mycorrhizal inoculant	White et al. 2006b

specific based on the data. Not only does phytoremediation potential vary among species, but it also varies with the variety of any one species (Table 2). Significant differences were found between varieties of *Cucurbita* sp. (squash and pumpkin) with up to an order magnitude difference in root:soil concentrations and two orders of magnitude for total plant uptake of *p,p*'-DDE (White 2002; White et al. 2003a). For two varieties of *Cucurbita pepo*, the translocation of DDE from soil to plant ranged from 0.4% to 2.4%, which approaches the values cited for "hyperaccumulating" plants used for heavy metal phytoremediation (White 2002). The success of *Cucurbita pepo* species (zucchini and pumpkin) in translocating DDT, DDD, and DDE is thought to result from high transpiration volumes, large aboveground biomass, and composition of root exudates (Lunney et al. 2004). Soil moisture and plant density also influence the uptake of *p,p*'-DDE from soil by *Cucurbita pepo* (Kelsey et al. 2006). Soil amendments such as biosurfactants (rhamnolipids produced by *Pseudomonas aeruginosa*) and mycorrhizal inoculation have both been shown to increase the translocation of *p,p*'-DDE in *Cucurbita pepo* (White et al. 2006a,b). Seven low molecular weight organic acids [succinic, tartaric, malic, malonic, oxalic, citric, and ethylenediaminetetraacetic acid (EDTA)] were tested for abiotic desorption of *p,p*'-DDE from soil with oxalic and citric acids being selected for use in conjunction with zucchini for phytoremediation. Both citric and oxalic acid had a positive influence on the translocation of *p,p*'-DDE from soil to shoot system, with the second crop showing even more of an increase in uptake than the first crop without further addition of the organic acids. It was suggested that the addition of low molecular weight organic acids disrupted the soil structure through chelation of inorganic ions, which enhanced the bioavailability of *p,p*'-DDE to *Cucurbita pepo* (White et al. 2003b).

Although much research has gone into phytoremediation by *Cucurbita pepo*, other species have shown potential as well. Suresh et al. (2005) found that hairy root cultures of *Chicorium intybus* (chicory) and *Brassica juncea* (brown mustard) took up ~87% of the ^{14}C-DDT within 24 hrs, with the transformation products being DDD, DDE, and DDMU (1-chloro-2,2-bis(4'-chlorophenyl)ethene), which suggests that these plants could be used for phytoremediation of DDT- and DDE-contaminated soil. Microflora such as algae and cyanobacteria have been reported to uptake DDT residues from soil, with the predominant degradation product being DDE for algae and DDD for the dinitrogen-fixating cyanobacteria (Megharaj et al. 1999, 2000). Aquatic species not only uptake but also transform DDT, with the order of efficacy being duckweed > elodea > parrot feather. The transformation was into DDD (31%–48%) and some DDE (4.6%–7.9%) being formed after 6d (Gao et al. 2000), in contrast to upland plants, where the major metabolite is usually DDE.

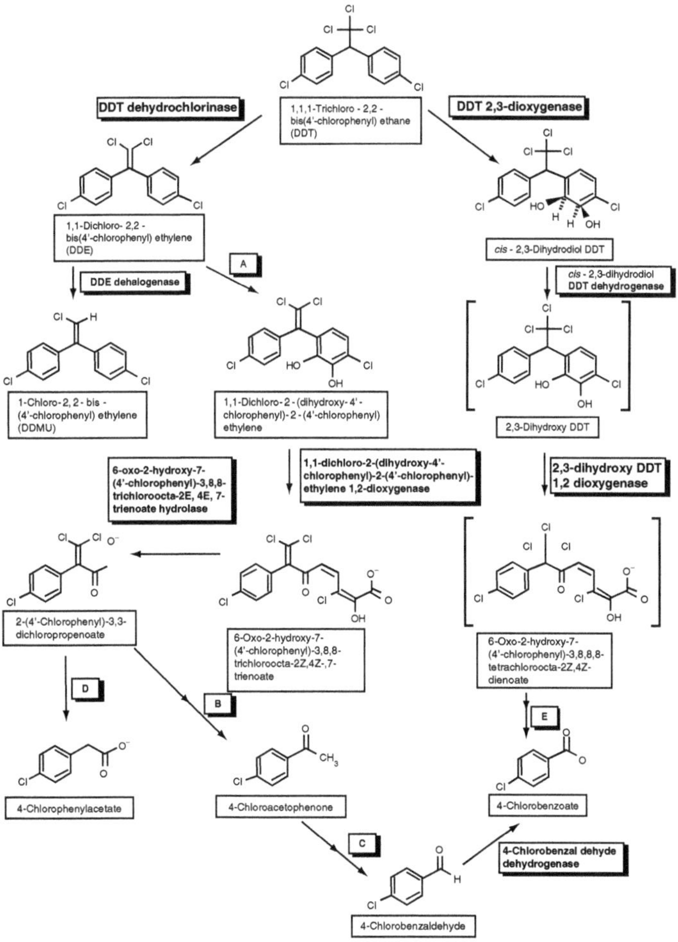

Fig. 1 Proposed degradation pathways for aerobic degradation of DDT and DDE from the UMBBD website (http://umbbd.msi.umn.edu/ddt/ddt_map.html). *Reproduced from Ellis LBM, Roe D, Wackett LP 2006, "The University of Minnesota Biocatalysis/Biodegradation Database: The First Decade," Nucleic Acids Research 34: D517-D521.*

III Aerobic Degradation and Remediation

The aerobic degradation pathway for DDT and DDE (Fig. 1), as presented by the University of Minnesota Biocatalysis/Biodegradation Database (Ellis et al. 2006), is one of several different pathways. Other variations have been suggested from several sources (Quensen et al. 1998; Hay and Focht 1998; Nadeau et al. 1998). The University of Minnesota Biocatalysis/Biodegradation Database (UMBBD) proposed degradation pathway is not fully verified; thus, each boxed letter and bracketed compound in Fig. 1 represent unknowns. To quote the UMBBD website: "The steps labeled **A, B, C, D, E** are mutiple steps whose intermediates are not identified yet. All descriptions as follows were proposed based on other similar experiments. In **A**, DDE is attacked by a dioxygenase at the *ortho* and *meta* positions. Such an attack would give rise to a 2,3-dihydrodiol-DDE intermediate. In **B** and **D**, 2-(4'-chlorophenyl)-3,3-dichloropropenoate may proceed via decarboxylation to yield 1,1-dichloro-(4'-chlorophenyl)ethane. The latter will undergo oxidation of the aliphatic side chain to yield 1,1-dichloro-(4'-chlorophenyl)ethanol which is further oxidized to yield 4-chloroacetophenone. The terminal ethyl group of 1,1-dichloro-(4'-chlorophenyl)ethane may also undergo oxidation to yield phenylacetic acid. In **C**, the transformation of 4-chloroacetophenone to 4-chlorobenzaldehyde may be via complete oxidation and subsequent decarboxylation of the terminal methyl group. In **E**, the ring-cleavage product would be further degraded to either a C-6- or C-5-chlorinated acid, depending on where the hydrolytic cleavage takes place."

Table 3 The five most effective microorganisms assessed in bioremediation studies and their associated reductions in DDE

Microorganism	Percent reduction of DDE (time)	Conditions	Reference
Terrabacter sp. (aerobic bacterium)	38 (10 d)	Soil-free aerobic liquid culture with biphenyl	Aislabie et al. 1999
Phanerochaete chrysoporium (fungus)	48 (18 d)	Nitrogen-limited liquid culture	Bumpus et al. 1993
Pleurotus plumonarius (fungus)	78 (60 d)	Sterile soil with sawdust compost	Gong et al. 2006
Alcaligenes denitrificans (facultative bacterium)	~26 (14 d)	Soil-free anaerobic liquid culture	Ahuja et al. 2001
Psuedomonas acidovorans (aerobic bacterium)	42 (25 d)	Soil-free aerobic liquid culture with biphenyl	Hay and Focht 1998

Although the proposed pathway for DDE appears to be fairly complete, there are very few reports of isolated bacterial or fungal cultures that can fully degrade it to CO_2 (Table 3). The pure aerobic cultures of *Pseudomonas acidovorans*, *Alicaligenes eutrophus*, and *Terrabacter* sp. that could degrade DDE came from liquid cultures that used the cometabolism of biphenyl to obtain the enzymes required (Aislabie et al. 1999; Hay and Focht 1998; Nadeau et al. 1998). Biphenyl-grown cells induced the production of biphenyl dioxygenase that catalyzed the degradation of DDE through meta-fission of the phenyl rings. However, not all cultures that are capable of producing biphenyl dioxygenase are capable of degrading DDE. The use of structural analogues to DDE, such as 4,4'-dichlorobiphenyl and 1,1-dichloroethylene, led Megharaj et al. (1997) to conclude that the recalcitrance of DDE to degradation by monooxygenase and biphenyl 2,3-dioxygenase enzymes produced by *Rhodococcus globerulus*, *Psuedomonas fluorescens*, *Mycobacterium vaccae*, and *Methylosinus trichosporium* may be the result of the 1,1-diphenylethenyl structure, rather than the extent of chlorination found in DDE. Other factors that may inhibit DDT and DDE degradation include metal content, such as elevated copper (Gaw et al. 2003) and arsenic (Van Zweiten et al. 2003) in the soil. The elevated arsenic levels resulted from its use in cattle-dipping vats as a tickicide. Arsenic was later replaced by DDT (Van Zweiten et al. 2003). The elevated copper levels found in orchard soils probably resulted from its use as a fungicide (Gaw et al. 2003).

Extracellular lignolytic enzymes produced by white rot fungi, *Phanerochaete chrysosporium* and *Pleurotus pulmonarius*, have been shown to be effective in degrading DDE (Bumpus et al. 1993; Gong et al. 2006). The *Phanerochaete chrysosporium* required nitrogen-limited cultures to effectively degrade DDT and DDE (Bumpus et al. 1993); however, *Pleurotus pulmonarius,* which secretes lignolytic enzymes under nitrogen-rich or -deficient conditions, degraded 78% of 10 mg DDE /kg soil within 5 wks (Gong et al. 2006). Wood-rotting basidomycetes are not the only type of fungi capable of degrading DDE. Genetically improved strains of *Fusarium solani* have been developed by parasexual hybridization from native fungi that slowly metabolize DDT, DDD, and DDE in soil (Mitra et al. 2001). Degradation by lignase enzymes from fungi can be inhibited by metal chelates such as EDTA and tetramethylethylenediamine (Aislabie et al. 1997).

IV Anaerobic Degradation and Remediation

Flooding soil can lead to anaerobic conditions, which have been shown, in some cases, to inhibit mineralization of DDT and DDE (Boul 1996; Xu et al. 1994). Until 1998, there was no convincing evidence to support reductive dechlorination of DDE (Quensen et al. 1998, 2001). The proposed reductive dechlorination pathway (Fig. 2) from DDE leads to DDMU (1-chloro-2,2-bis (p-chlorophenol)ethene).

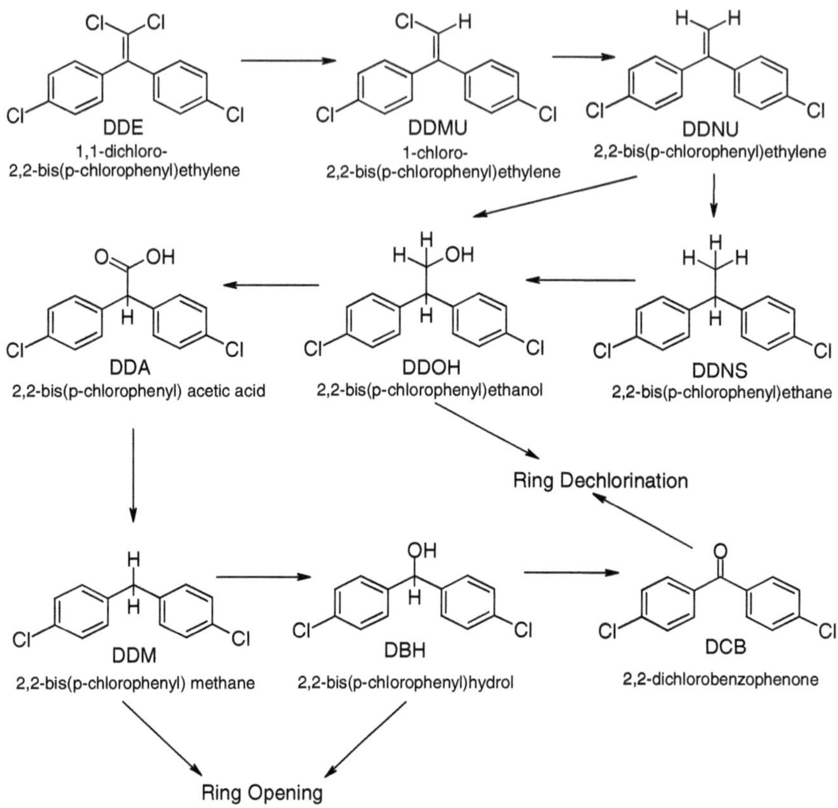

Fig. 2 Proposed degradation pathways for anaerobic degradation of DDE. *Reprinted from Chemosphere, Vol. 62, T. Eggen and A. Majcherczyk, Effects of zero-valent iron and temperature on the transformation of DDT and its metabolites in lake sediment, 1116-1125, with kind permission from Elsevier, 2006.*

Laboratory experiments have demonstrated that the conversion of DDE to DDMU has occurred under methanogenic and sulfidogenic conditions, although the presence of sulfate and low temperatures did lower the rate of degradation (Quensen et al. 1998, 2001). The biotransformation of DDE in anoxic sediments has been confirmed to occur in at least one other site (Huang et al. 2001). Batch reactor experiments using alternating aerobic and anaerobic conditions found very little DDE degradation after 105 d (Strompl and Thiele 1997). However, a batch reactor using a mixed culture with surfactants, Triton X-114 or Brij 35, in conjunction with reducing agents, Na_2S or cysteine HCl, did reduce DDE concentration, although not as much as DDT or DDD was degraded. It was assumed that the greater degradation was caused by the surfactant solubilizing the DDE, making it more bioavailable. The lower amount of DDE degradation than that observed for DDT and DDD was assumed to be because DDE binds more strongly to the soil particles (You et al. 1996). It should be noted that in anaerobic microcosm experiments with cellulose and the surfactant Brij 30, using aged DDT-contaminated soil that initially contained DDE, there was no change in the concentration of DDE after 31 wks, although DDT degradation was accelerated and DDD concentration increased (Walters and Aitken 2001). Similarly, in other microcosm experiments where Na^+ was added to aged DDT-contaminated soils in an effort to increase clay dispersal, which leads to more dissolved organic carbon and greater bioavailability of the sorbed contaminants, it was reported that the DDT degradation increased, DDD accumulation increased, but DDE remained relatively the same throughout the experiments. The conditions used in these experiments gave anoxic levels similar to those found for denitrification (Kantachote et al. 2004). The only literature located that described a pure culture to degrade DDE anaerobically involved a denitrifier, *Alcaligenes denitrificans* (Ahuja et al. 2001). Degradation was accelerated under glucose but was inhibited by sodium acetate and sodium succinate. The addition of biphenyl fumes had no effect on the rate of DDE disappearance. Denitrifying conditions can be easily reached by flooding the soil (Kantachote et al. 2004).

V Abiotic Remediation and Degradation

The effect of soil flooding on the binding of DDT and DDE was examined using microcosm experiments. Using ^{14}C -DDT, Boul (1996) found that in nonflooded conditions 6.7% and 9.7% of DDT and DDE, respectively, were bound to the soil over a 42-d period. Under flooded conditions, the amounts increased to 24.5% DDT and 11.5% DDE. It was also reported that <0.7% of the ^{14}C was emitted as $^{14}CO_2$ under nonflooded conditions and that virtually no $^{14}CO_2$ evolved when the soil was flooded. Flooding, as a land management practice, along with deep plowing, were used in a 23-yr experiment to determine the effects on the contaminants by these practices (Spencer et al. 1996). It was found that the major residue was *p,p*'-DDE, with the greatest amounts being in the nonflooded and deep-plowed plots; deep plowing possibly reduced the amount of DDT and DDE volatilization. Irrigation significantly enhanced volatilization of DDT residues, particularly *p,p*'-DDE. These findings are consistent with the earlier report that long-term irrigation with superphosphate fertilizer lowered *p,p*'-DDE residues compared to nonirrigated fields. The irrigation did not affect the DDT residue distribution by depth, which indicated that irrigation did not cause an increase in the leaching potential (Boul et al. 1994).

To increase the degradation rates of DDT, DDD, and DDE in flooded soils and sediments, some researchers have used Fe°, zero-valent iron (Eggen and Majcherczyk 2006; Pirnie at al. 2006; Satapanajaru et al. 2006; Sayles et al. 1997; Yao et al. 2006). Sayles et al. (1997) used a soil-free system and found that the rate of dechlorination of DDT and DDE was independent of the iron powder concentration, but that the rate was much higher when the surfactant Triton X-114 was used in conjunction with the iron powder. Various combinations of calcium peroxide, zero-valent iron, iron sulfide, and hydrogen peroxide were tested in aqueous solutions with and without the surfactant Triton X-114. Although these systems did degrade

DDT, a large amount of DDE was generated (Pirnie et al. 2006). Eggen and Majcherczyk (2006) used DDT contaminated and naturally aged sediments with zero-valent iron at two different temperatures, 9°C and 22°C, under anoxic conditions. Under these conditions, some degradation of DDE occurred in the initial 10- to 20-week incubation time at 22°C then remained constant up to 40 weeks, whereas there was no transformation at 9°C. The difference in DDE degradation between the microcosms with and without Fe° was not significant. Using acidic rice paddy soil spiked with *p,p*'-DDT, Yao et al. (2006) reported that Fe° or Fe° + $Al_2(SO_4)_3$ exhibited an initial 25%–35% decrease in *p,p*'-DDE in the first week, a return to the initial concentration after 2 wk, and, finally, a dramatic decrease in the following 3 wk with the final concentration of *p,p*'-DDE at 60% of the original. However, the effect of these adjuncts was complicated by the observation that the control removed more extractable *p,p*'-DDE than the treated samples.

In contrast to the Fe° systems, which can take days to weeks and leave residual DDE, the complete removal of DDT within 10 min was observed for an acidified aqueous system treated with palladium/magnesium (Mg°/Pd^{+4}) particles under ambient temperatures and pressures with no DDE residues detected (Engelmann et al. 2001). Palladized magnesium has also been used to dechlorinate more than 99% extractable DDT from soil as well as 88% of DDE from a soil slurry made from 1 g soil spiked with 50 μg DDE and aged 30 d (Gautam and Suresh 2006). More complex supported-catalyst systems have been developed involving Pd/C, Pt/C, or the more inexpensive Raney-Ni catalyst (1:1 Ni/Al alloy) to degrade DDT and its metabolites (Zinovyev et al. 2005). The complexity results from the two-phase organic/aqueous liquid system, which requires a quartenary ammonium salt and KOH to act as a promoter/carrier. When KOH and a quartenary ammonium salt are present, DDE is rapidly dechlorinated and the ethylene double bond is reduced. Another Pd/C catalyst system that can rapidly degrade DDE involves the addition of triethylamine and hydrogen under ambient pressure and temperature. The authors claim that this catalyst system is simple, effective, reliable, and inexpensive (Monguchi et al. 2006).

Titanium dioxide has been used as a catalyst in the degradation of DDT, DDD, and DDE by UV light in soil (Quan et al. 2005). The photodegradation rate increased with an increase in pH and photon flux rate, but decreased with an increase in humic acid content of the soil. It was hypothesized that the humic acid either reduced the amount of light reaching the TiO_2 or that the humic acid quenched the radicals responsible for oxidizing the contaminants. *p,p*'-DDE, *p,p*'-DDD, and DDMU were all reported to be degradation products, but all compounds were further degraded by TiO_2 and UV light. DDT can be degraded by UV light without TiO_2 in aqueous system that contains a surfactant such as Brij 52; however, DDE and DDD are the resulting products (Chu 1999). DDE was found to be degraded in a 1:1 acetonitrile:water system with DDT, methylene green (photosensitizer), and triethylamine (electron donor) under visible light, but it was not degraded if the concentration of the methylene green fell below 10^{-7} M (Lin and Chang 2007). Overall, the abiotic methods have had varying degrees of success in the remediation and degradation of DDE (Table 4).

Table 4 The efficacies of selected abiotic treatments assessed in soil remediation studies and their associated reductions in DDE

Treatment	Percent reduction of DDE (time)	Conditions	Reference
Deep plowing	14.2 (23 yr)	No flooding No amendments	Spencer et al. 1996
Deep plowing and flooding	69.6 (23 yr)	45 ton ha^{-1} manure	Spencer et al. 1996
Shallow flooding	11.5 (42 d)	59% soil water content	Boul 1996
Photodegradation	~80 (12 hr)	1% TiO_2 on soil with UV light	Quan et al. 2005
Catalysis	>99 (24 hr)	Soil slurry Biosurfactant Pd/Mg catalyst	Gautam and Suresh 2006
Catalysis	~60 (7 wk)	Acidic rice paddy soil Fe° catalyst	Yao et al. 2006
Catalysis	~70 (40 wk)	Lake sediment at 22°C Fe° catalyst	Eggen and Majcherczyk 2006

Summary

DDT and its metabolites, DDD and DDE, have been shown to be recalcitrant to degradation. The parent compound, DDT, was used extensively worldwide starting in 1939 and was banned in the United States in 1973. The daughter compound, DDE, may result from aerobic degradation, abiotic dehydrochlorination, or photochemical decomposition. DDE has also occurred as a contaminant in commercial-grade DDT. The *p,p*'-DDE isomer is more biologically active than the *o,p*-DDE, with a reported half-life of ~5.7 years. However, when DDT was repeatedly applied to the soil, the DDE concentration may remain unchanged for more than 20 yr.

Remediation of DDE-contaminated soil and water may be done by several techniques. Phytoremediation involves translocating DDT, DDD, and DDE from the soil into the plant, although some aquatic species (duckweed > elodea > parrot feather) can transform DDT into predominantly DDD with some DDE being formed. Of all the plants that can uptake DDE, *Cucurbita pepo* has been the most extensively studied, with translocation values approaching "hyperaccumulation" levels. Soil moisture, temperature, and plant density have all been documented as important factors in the uptake of DDE by *Cucurbita pepo*. Uptake may also be influenced positively by amendments such as biosurfactants, mycorrhizal inoculants, and low molecular weight organic acids (e.g., citric and oxalic acids).

DDE microbial degradation by dehalogenases, dioxygenases, and hydrolases occurs under the proper conditions. Although several aerobic degradation pathways have been proposed, none has been fully verified. Very few aerobic pure cultures are capable of fully degrading DDE to CO_2. Cometabolism of DDE by

Pseudomonas sp., *Alicaligens* sp., and *Terrabacter* sp. grown on biphenyl has been reported; however, not all bacterial species that produce biphenyl dioxygenase degraded DDE. Arsenic and copper inhibit DDE degradation by aerobic microorganisms. Similarly, metal chelates such as EDTA inhibit the breakdown of DDE by the extracellular lignolytic enzymes produced by white rot fungi. The addition of adjutants such as sodium ion, surfactants, and cellulose increased the rate of DDT aerobic or anaerobic degradation but did little to enhance the rate of DDE disappearance under anaerobic conditions. Only in the past decade has it been demonstrated that DDE can undergo reductive dechlorination under methanogenic and sulfidogenic conditions to form the degradation product DDMU, 1-chloro-2,2'-bis-(4'-chlorophenyl)ethane. The only pure culture reported to degrade DDE under anaerobic conditions was the denitrifier *Alcaligens denitrificans*. The degradation of DDE by this bacterium was enhanced by glucose, whereas biphenyl fumes had no effect.

Abiotic remediation by DDE volatilization was enhanced by flooding and irrigation and deepplowing inhibited the volatilization. The use of zero-valent iron and surfactants in flooded soils enhanced DDT degradation but did not significantly alter the rate of DDE removal. Other catalysts (palladized magnesium, palladium on carbon, and nickel/aluminum alloys) degraded DDT and its metabolites, including DDE. However, these systems are often biphasic or involve explosive gases or both. Safer abiotic alternatives use UV light with titanium oxide or visible light with methylene green to degrade DDT, DDD, and DDE in aqueous or mixed solvent systems.

Remediation and degradation of DDE in soil and water by phytoextraction, aerobic and anaerobic microorganisms, or abiotic methods can be accomplished. However, success has been limited, and great care must be taken that the method does not transfer the contaminants to another locale (by volatilization, deep plowing, erosion, or runoff) or to another species (by ingestion of accumulating plants or contaminated water). Although the remediation of DDT-, DDD-, and DDE-contaminated soil and water is beset with myriad problems, there remain many open avenues of research.

References

Ahuja R, Awasthi N, Manickam N, Kumar A (2001) Metabolism of 1,1-dichloro-2,2-bis(4-chlorophenyl)ethylene by *Alicaligenes denitrificans*. Biotech Lett 23:423–426.

Aigner E, Leone A, Falconer R (1998) Concentrations and enantiomeric ratios of organochlorine pesticides in soil from the U.S. corn belt. Environ Sci Technol 32:1162–1168.

Aislabie J, Richards N, Boul H (1997) Microbial degradation of DDT and its residues: a review. N Z J Agric Res 40:269–282.

Aislabie J, Davison A, Boul H, Franzmann P, Jardine D, Karuso P (1999) Isolation of *Terrabacter* sp. DDE-1, which metabolizes 1,1-dichloro-2,2-bis(4-chlorophenyl)ethylene when induced by biphenyl. Appl Environ Microbiol 65:5607–5611.

Atlas E, Foster R, Glam C (1982) Air-sea exchange of high molecular weight organic pollutants: laboratory studies. Environ Sci Technol 16:283–286.

ATSDR (2002) Toxicological profile for DDT, DDE, and DDD. U.S. Dept. of Health and Human Services, Agency for Toxic Substance and Disease Registry, Division of Toxicology/ Toxicology Information Branch, Atlanta, GA. Available on-line at http://www.atsdr.cdc.gov/ toxprofiles/tp35.pdf (last checked on 3/20/2007).

Beyer W, Krynitsky A (1989) Long-term persistence of dieldrin, DDT, and heptachlor epoxide in earthworms. Ambio 18:271–273.

Boul H (1996) Effects of soil moisture on the fate of DDT and DDE *in vitro*. Chemosphere 32:855–866.

Boul H, Garnham M, Hucker D, Baird D, Aislabie J (1994) The influence of agricultural practices on the levels of DDT and its residues in soil. Environ Sci Technol 28:1397–1402.

Bumpus J, Powers R, Sun T (1993) Biodegradation of DDE (1,1-dichloro-2,2-bis(4-chlorophenyl) ethane) by *Phanerochaete chrysosporium*. Mycol Res 97:95–98.

Chu W (1999) Photodechlorination mechanism in a UV/surfactant system. Environ Sci Technol 33:421–425.

Cliath M, Spencer W (1972) Dissipation of pesticides from soil by volatilization of degradation products. I. Lindane and DDT. Environ Sci Technol 6:910–914.

Cortes D, Hites R (2000) Detection of statistically significant trends in atmospheric concentrations of semivolatile compounds. Environ Sci Technol 34:2826–2829.

Ding J, Wu S (1997) Transport of organochlorine pesticides in soil columns enhanced by dissolved organic carbon. Water Sci Technol 35:139–145.

Eggen T, Majcherczyk A (2006) Effects of zero-valent iron (Fe°) and temperature on the transformation of DDT and its metabolites in lake sediment. Chemosphere 62:1116–1125.

Ellis L, Roe D, Wackett L (2006) The University of Minnesota biocatalysis/biodegradation database: the first decade. Nucleic Acids Res 34:D517–D521.

Engelmann M, Doyle J, Cheng I (2001) The complete dechlorination of DDT by magnesium/palladium bimetallic particles. Chemosphere 41:195–198.

Gao J, Garrison A, Hoehamer C, Mazur C, Wolfe N (2000) Uptake and phytotransformation of *o,p*'-DDT and *p,p*'-DDT by axenically cultivated aquatic plants. J Agric Food Chem 48:6121–6127.

Gautam S, Suresh S (2006) Dechlorination of DDT, DDD and DDE in soil (slurry) phase using magnesium/palladium system. Colloid Interface Sci 304:144–151.

Gaw S, Palmer G, Kim N, Wilkins A (2003) Preliminary evidence that copper inhibits the degradation of DDT to DDE in pip and stonefruit orchard soils in Auckland region, New Zealand. Environ Pollut 122:1–5.

Gonzalez M, Miglioranza K, De Moreno J, Moreno V (2003) Organochlorine pesticide residues in leek (*Allium porrum*) crops grown on untreated soils from agricultural environment. J Agric Food Chem 51:5024–5029.

Gong J, Chan K, Chiu S (2006) Remediation of persistent toxic substances: toxicities of DDE on wheat and bioremediation of DDE by fungus *Pleurotus pulmonarius*. Hum Ecol Risk Assess 12:221–235.

Harner T, Bidleman T, Jantunen L, Mackay D (2001) Soil-air exchange model of persistent pesticides in the United States cotton belt. Environ Toxicol Chem 20:1612–1621.

Hay A, Focht D (1998) Cometabolism of 1,1-dichloro-2,2-bis(4-chlorophenyl)ethylene by *Pseudomonas acidovorans* M3GY grown on biphenyl. Appl Environ Microbiol 64:2141–2146.

Heberer T, Dunnbier U (1999) DDT metabolite bis(chlorophenyl)acetic acid: the neglected environmental contaminant. Environ Sci Technol 33:2346–2351.

Huang H, Liu S, Kuo C (2001) Anaerobic biodegradation of DDT residues (DDT, DDD, and DDE) in estuarine sediment. J Environ Sci Health Part B 36:273–288.

Hung H, Halsall C, Blanchard P, Li H, Fellin P, Stern G, Rosenberg B (2002) Temporal trends of organochlorine pesticides in the Canadian arctic atmosphere. Environ Sci Technol 36:862–868.

Hussain A, Maqbool U, Asi M (1994) Studies on the dissipation and degradation of C-14 DDT and C-14 DDE in Pakastani soils under field conditions. J Environ Sci Health Part B 29:1–15.

Kantachote D, Singleton J, Naidu R, McClure N, Megharaj M (2004) Sodium application enhances DDT transformation in a long-term contaminated soil. Water Air Soil Pollut 154:115–124.

Karickhoff S (1981) Semiempirical estimation of hydrophobic pollutants on natural sediments and Soils. Chemosphere 10:833–846.

Kelsey J, Colino A, Koberle M, White J (2006) Growth conditions impact 2,2-bis(*p*-chlorophenyl)-1,1-dichloroethylene (*p,p*'-DDE) accumulation by *Cucurbita pepo*. Int J Phytoremed 8:261–271.

Lin C, Chang T (2007) Photosensitized reduction of DDT using visible light: the intermediates and pathways of dechlorination. Chemosphere 66:1003–1011.

Lunney A, Zeer B, Reimer K (2004) Uptake of weathered DDT in vascular plants: potential for phytoremediation. Environ Sci Technol 38:6147–6154.

Masters P, Inman D (2000) Transport and fate of organochlorines discharged to the salt marsh at upper Newport Bay, California, USA. Environ Toxicol Chem 19:2076–2084.

Megharaj M, Jovcic A, Boul J, Thiele J (1997) Recalcitrance of 1,1-dichloro-2,2-bis (*p*-chlorophenyl)ethylene (DDE) to cometabolic degradation by pure cultures of aerobic and anaerobic bacteria. Arch Environ Contam Toxicol 33:141–146.

Megharaj M, Boul H, Thiele J (1999) Effects of DDT and its metabolites on soil algae and enzymatic activity. Biol Fertil Soils 29:130–134.

Megharaj M, Kantachote D, Singleton I, Naidu R (2000) Effects of long-term contamination of DDT on soil microflora with special reference to soil algae and algal transformation of DDT. Environ Pollut 109:35–42.

Metcalf R (1995) Insect control technology. In: Kroschwitz, J, Howe-Grant, M (eds) Kirk-Othmer Encyclopedia of Chemical Technology, vol 14. Wiley, New York, pp 524–602.

Mitra J, Mukherjee P, Kale S, Murthy N (2001) Bioremediation of DDT in soil by genetically improved strains of soil fungus *Fusarium solani*. Biodegradation 12:235–245.

Monguchi Y, Kume A, Sajiki H (2006) Facile and catalytic degradation method of DDT using Pd/C-Et_3N system under ambient pressure and temperature. Tetrahedron 62:8384–8392.

Nadeau L, Menn F, Breen A, Sayler G (1998) Aerobic degradation of 1,1,1-trichloro-2,2-bis(4-chlorophenyl)ethane (DDT) by *Alcaligenes eutrophus* A5. Appl Environ Microbiol 60:51–55.

Pirnie E, Talley J, Hundai L (2006) Transformation of DDT and its metabolites by various abiotic methods. J Environ Eng 132:560–564.

Quan X, Zhao X, Chen S, Zhao H, Chen J, Zhao Y (2005) Enhancement of *p,p*'-DDT photodegradation on soil surfaces using TiO_2 induced by UV-light. Chemosphere 60:266–273.

Quensen J, Mueller S, Jain M, Tiedje J (1998) Reductive dechlorination of DDE to DDMU in marine sediment microcosms. Science 280:724–724.

Quensen J, Tiedje J, Jain M, Mueller S (2001) Factors controlling the rate of DDE dechlorination to DDMU in Palos Verdes margin sediments under anaerobic conditions. Environ Sci Technol 35:286–291.

Satapanajaru T, Anurakpongsatorn F, Songasen A, Boparai H, Park J (2006) Using low-cost iron byproducts from automotive manufacturing to remediate DDT. Water Air Soil Pollut 175:361–374.

Sayles G, You G, Wang M, Kupferle M (1997) DDT, DDD, and DDE dechlorination by zero-valent iron. Environ Sci Technol 31:3448–3454.

Sharpe R (1995) Reproductive biology: another DDT connection. Nature (Lond) 375:538–539.

Shen L, Wania F (2005) Compilation, evaluation, and selection of physical-chemical property data for organochlorine pesticides. J Chem Eng Data 50:742–768.

Spencer W, Singh G, Taylo RA, LeMert A, Cliath M, Farmer W (1996) DDT persistence and volatility as affected by management practices after 23 years. J Environ Qual 35:815–821.

Strompl C, Thiele J (1997) Comparative fate of 1,1-diphenylethylene (DPE), 1,1-dichloro-2,2-bis(4-chlorophenyl)-ethylene (DDE), and pentachlorophenol (PCP) under alternating aerobic and anaerobic conditions. Arch Environ Contam Toxicol 33:350–356.

Suresh B, Sherkhane P, Kale S, Eapen S, Ravishankar G (2005) Uptake and degradation of DDT by hairy root cultures of *Chicorium intybus* and *Brassica juncea*. Chemosphere 61:1288–1292.

Thiele J, Boul H, Jovcic A, Megharaj M (1997) Recalcitrance of 1,1-dichloro-2,2-bis(*p*-chlorophenol)ethylene (DDE) to cometabolic degradation by pure cultures of aerobic and anaerobic bacteria. Arch Environ Contam Toxicol 33:141–146.

Turusov V, Rakitsky V, Tomatis L (2002) Dichlorodiphenyltrichloroethane (DDT): ubiquity, persistence, and risks. Environ Health Perspect 110:125–128.

USGS (1999) The quality of our nation's waters: nutrients and pesticides. Circular 1225. U.S. Department of the Interior, U.S. Geological Survey, Reston, VA. Available online at http://pubs.usgs.gov/circ/circ1225/ (last checked 3/29/07).

Van Zweiten L, Ayres M, Morris S (2003) Influence of arsenic co-contamination on DDT breakdown and microbial activity. Environ Pollut 124:331–339.

Walters G, Aitken M (2001) Surfactant-enhanced solubilization and anaerobic biodegradation of 1,1,1-trichloro-2,2-bis(*p*-chlorophenyl)-ethane (DDT) in contaminated soil. Water Environ Res 73:15–23.

White J (2002) Differential bioavailability of field-weathered *p,p*'-DDE to plants of the *Cucurbita* and *Cucumis* genera. Chemosphere 49:143–152.

White J, Wang X, Gent M, Iannucci-Berger W, Eitzer B, Schultes N, Artenzo M, Mattina M (2003a) Subspecies-level variation in the phytoextraction of weathered *p,p*'-DDE by *Cucurbita pepo*. Environ Sci Technol 37:4368–4373.

White J, Mattina M, Lee W, Eitzer B, Iannucci-Berger W (2003b) Role of organic acids in enhancing the desorption and uptake of weathered *p,p*'-DDE by *Cucurbita pepo*. Environ Pollut 124:71–80.

White J, Parrish Z, Isleyen M, Gent M, Ianmucci-Berger W, Eitzer B, Mattina M (2005) Uptake of weathered *p,p*'-DDE by plant species effective at accumulating soil elements. Microchem J 81:148–155.

White J, Ross D, Gent M, Eitzer B, Mattina M (2006a) Effect of mycorrhizal fungi on the phytoextraction of weathered *p,p*'-DDE by *Cucurbita pepo*. J Hazard Mater B137:1750–1757.

White J, Parrish Z, Gent M, Iannucci-Berger W, Eitzer B, Isleyen M, Mattina M (2006b) Soil amendments, plant age, and intercropping impact of p,p'-DDE bioavailability to *Cucurbita pepo*. J Environ Qual 35:992–1000.

Xu B, Gang J, Zhang Y, Lin H (1994) Behavior of DDT in Chinese tropical soils. J Environ Sci Health Part B 29:37–46.

Yao F, Jiang X, YuG, Wang F, Bian Y (2006) Evaluation of accelerated dechlorination of *p,p*'-DDT in acidic paddy soil. Chemosphere 64:628–633.

You G, Sayles G, Kupferle M, Kim I, Bishop P (1996) Anaerobic DDT biotransformation: enhancement by application of surfactants and low oxidation reduction potential. Chemosphere 32:2269–2284.

Zinovyev S, Shinkova N, Perosa A, Tundo P (2005) Liquid phase hydrodechlorination of dieldrin and DDT over Pd/C and Raney-Ni. Appl Catal B Environ 55:39–48.

Surfactant Effects on Environmental Behavior of Pesticides

Toshiyuki Katagi[1]

I Introduction

Pesticides having a wide variety of structures have been and are being developed to exhibit an intended efficacy to pests, fungal diseases, and weeds, and are usually applied as formulations to these targets in the field. Many types of formulations have

---

[1]Sumitomo Chemical Co., Ltd. Environmental Health Science Laboratory, 2-1 Takatsukasa 4-Chome, Takarazuka, Hyogo 665-8555, Japan.

Communicated by G.W. Ware.

been developed by considering the physicochemical properties of the pesticide, not only to maximize their efficacy but also to prevent unfavorable environmental contamination from pesticide and its degradation products. A lower water solubility of the pesticide usually leads to utilization of a water-miscible organic solvent and formulation additives (adjuvants) including surface-active agents (surfactants) to solubilize, suspend, or disperse the active ingredient (a.i.) of pesticide in its aqueous solution being applied in agricultural practices. Among adjuvants, surfactants are some of the most important components and can improve the biological activity by modifying spray droplet size, retention, and spreading on leaf surfaces or by enhancing uptake and translocation of the a.i. to crop (Knowles 2001). The various kinds of surfactants with a wide variety of molecular weight and ionic character as well as natural and petroleum oils are used as wetting, spreading, sticking, and penetration agents and humectants (Hazen 2000; Krogh et al. 2003).

The main types of pesticide formulation containing surfactant are listed in Table 1. The surfactant in these formulations generally occupies less than 10% (w/w), and the remaining consists of various additives such as oils, polymers, and carriers. The emulsifiable concentrate (EC) using petroleum-based solvents and dust (D) made of inorganic carriers are typical formulations that have been used, but now they are being replaced by more environmentally benign formulations to avoid unfavorable ecotoxicity and contamination by spray drift.

When a pesticide formulation is applied in the field, the adjuvants including surfactants are distributed via many routes in the environment, depending on their physicochemical properties (Fig. 1). Surfactants also undergo abiotic degradation such as hydrolysis and sunlight photolysis together with metabolic transformation

Table 1 Principal types of pesticide formulation containing surfactants in their typical composition

Formulation type	A.I. %	Surfactant %	Other components (%)
Granules (G)	1–40	0–5	Stabilizer (1–2), polymer or resin (0–10), Binder (0–5), carrier (to 100)
Wettable powder (WP)	10–80	1–2	Dispersing agent (2–5), antifoaming agent (0.1–1) Inert filler/carrier (to 100)
Soluble concentrate (SL)	20–70	5–15	Antifreeze agent (5–10), water-miscible solvent (to 100)
Emulsifiable concentrate (EC)	20–70	5–10	Solvent/co-solvent (to 100)
Suspension concentrate (SC)	20–70	2–5	Propylene glycol antifreeze (5–10), antisettling Agent (0.2–2), water (to 100)
Water-dispersible granules (WDG)	50–90	1–3	Dispersing agent (3–15), disintegrating agent (0–15) Soluble or insoluble filler (to 100)
Oil-in-water emulsion (EW)	5–30	<5%	Stabilizer, thickener

Source: From Knowles (2001) and Copping (2000).

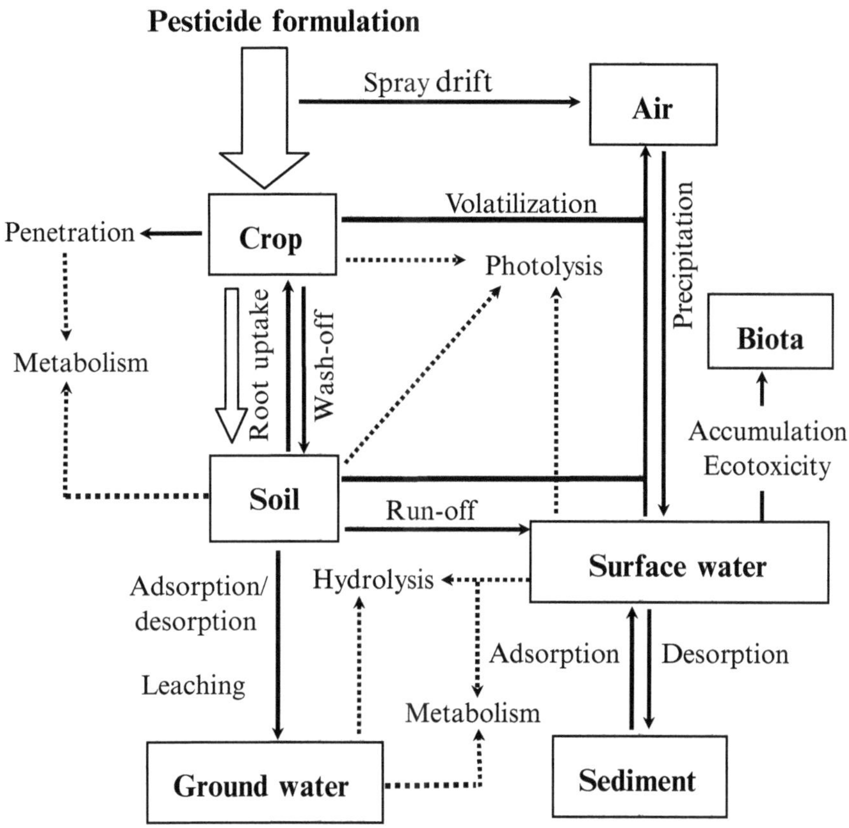

Fig. 1 Distribution and transformation of pesticide and adjuvants after application in the environment

in crops and soil. To know the environmental behavior of surfactants, each process in Fig. 1 has been extensively studied and monitored by many researchers together with their ecotoxicological impacts. In addition to spray drift, runoff events would facilitate their distribution especially in water. Pedersen et al. (2003) have identified by gas chromatography-mass spectrometry (GC-MS) many petroleum distillates and degradates of nonionic surfactants originating from pesticide formulations. The occurrence of many surfactants and their degradates originating from agricultural, laundry, and cosmetic uses has also been reported for sewage effluents, treated sludges, natural water, and associated sediments (Eichhorn 2003; Knepper et al. 2003; Krogh et al. 2003; Sanderson et al. 2006; Ying 2006). The biodegradable surfactants used in pesticide formulations together with their metabolites have been scarcely detected at the total residual level exceeding around 10 ppb in most natural surface waters and associated sediments, but more residues were detected for linear alkylbenzene sulfonates. Thus, the high concentration of surfactants in the area of pesticide application is generally reduced to a trace level through distribution to other areas such as a body of water via dilution and degradation processes.

Through the registration process of a new pesticide, many physicochemical properties and environmental fate profiles should be clarified from laboratory studies basically using an a.i., followed by field trials using a typical formulation product to confirm the actual residues of a pesticide and its relevant metabolites. Because surfactants above the characteristic critical micelle concentration (cmc) aggregate to micelles that result in a different reaction environment from an aqueous phase, basic physical and chemical processes for a pesticide are most likely to be modified in formulation (Bunton and Savelli 1986; Edwards et al. 1992; Fendler and Fendler 1975; Grieser and Drummond 1988; Taşcioğlu 1996; Thomas 1980). Furthermore, some interactions with surfactant monomers are known to result in different behavior of a substrate from those in an aqueous phase. From this point of view, the physicochemical properties should be measured for an a.i., and the usage of a formulated product is not recommended in the Organization for Economic Development (OECD) testing guidelines for hydrolysis (OECD 2004), soil adsorption-desorption (OECD 2000), and soil metabolism (OECD 2002a,b) studies. Although some of the underlying mechanisms of the surfactant effects on each process in Fig. 1 have been investigated for small organic molecules and polycyclic aromatic hydrocarbons (PAH), their effects on environmental behavior of pesticides have not been systematically examined. Based on these considerations, this review first deals with basic properties of surfactants used in pesticide formulation together with their environmental behavior. The effects on the behavior of pesticide are then discussed for the typical formulations, including surfactants, after the basic investigations on small organic molecules together with its mechanism.

II Surfactants in Pesticide Formulations

A *Type and Properties of Surfactant*

The surfactant molecule basically possesses a hydrophilic polar head group and a hydrophobic part such as a long alkyl chain, thus showing an amphiphilic character. Surfactants mostly lower the surface tension of an air–water interface from 72 mN/m to around 30–40 mN/m and are generally classified from their chemical structures mainly in three categories: cationic, anionic, and nonionic. The chemical structures of surfactants typically used for pesticide formulation are listed in Fig. 2. Nonionic surfactants with a polyethoxy chain as a hydrophilic part are the most popular. The commercially available nonionic alcohol and alkylphenoxy ethoxylates are usually polydisperse with the carbon numbers of an alkyl chain and CH_2CH_2O segment being distributed around their averages. The most familiar cationic surfactant is a quaternary ammonium salt such as hexadecyltrimethylammonium bromide (HDTMA Br) having the $(CH_3)_3N^+CH_2$ moiety as a polar head group and used as a thickener for suspension and emulsion formulations. In various types of anionic surfactants, either a sulfate or sulfonate group exhibits a hydrophilic character and the hydrophobic part is usually a long saturated alkyl chain or

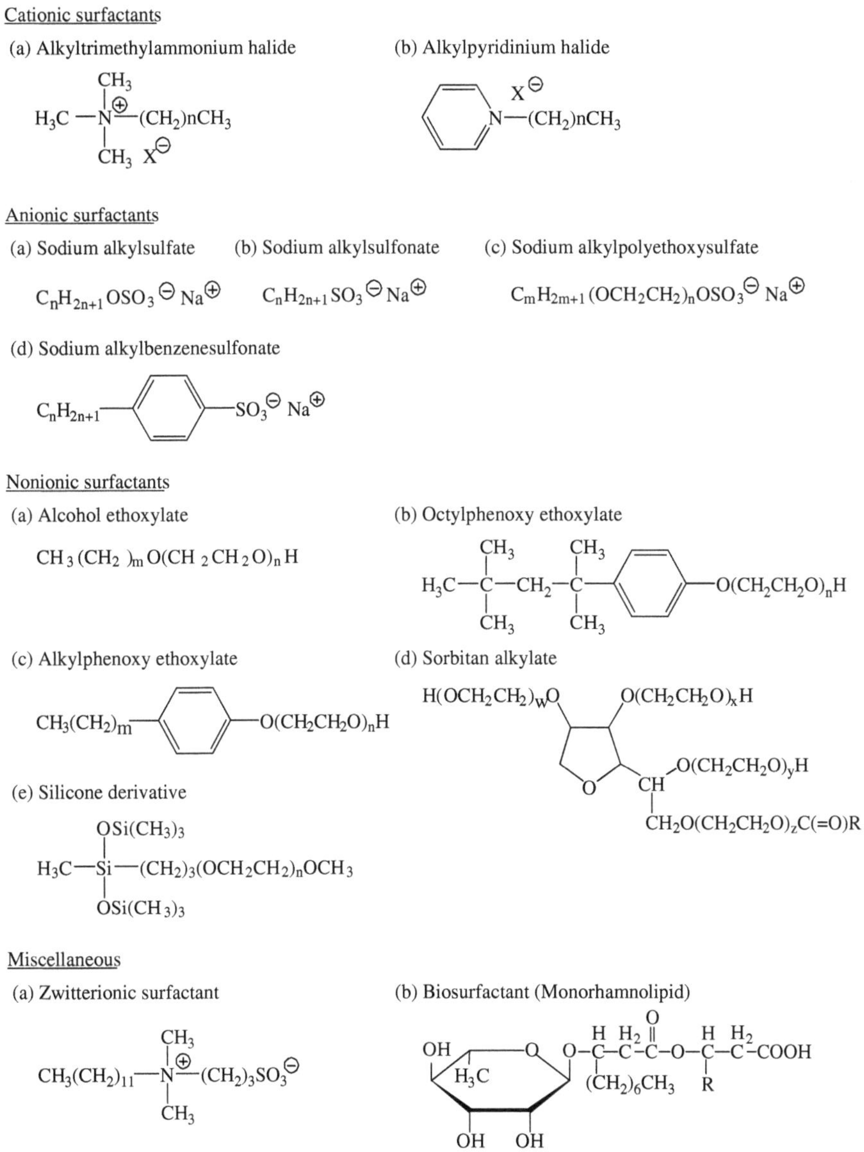

Fig. 2 Chemical structures of typical surfactants

alkylbenzene. Recently, silicone derivatives such as Silwet L-77 showing a higher surface activity have been developed for formulations (Stevens 1993). As other categories, a zwitterionic surfactant having both cationic and anionic parts in a molecule and many kinds of biosurfactants including rhamnolipids produced by many microorganisms (Makkar and Rockne 2003) are known. The dirhamnolipid has been demonstrated to form multilamellar vesicles in a heterogeneous size above its cmc instead of spherical micelles (Sánchez et al. 2007). To grasp the hydrophobicity or hydrophilicity of a surfactant, a hydrophilic–lipophilic balance (HLB) shown below has been conveniently developed by Griffin (1949) and is widely used:

$$\mathrm{HLB} = 20(M_h/M)$$

where M_h and M are the molecular masses of a hydrophilic part and the whole molecule, respectively. The lower and higher values, respectively, mean that surfactant is hydrophobic and hydrophilic. According to this index, it is generally considered that surfactants having the HLB values of 2–7 and 7–18 are adequate for preparing emulsifiable concentrate (EC) and oil-in-water emulsion (EW) formulations (Knowles 2001).

A surfactant molecule is dissolved in water as a monomer at a lower concentration but tends to aggregate to form micelles at cmc (Fendler and Fendler 1975; Bunton et al. 1991; Ying 2006). Micelles are dynamic colloidal aggregates and optically transparent but scatter light. The size and shape are dependent on the concentration of surfactants and the solution composition. At the concentration slightly above cmc, the spherical micelles with a typical radius of 10–30 Å are formed by specific numbers of surfactant molecules (aggregation number) (Fig. 3a), but either increase of surfactant concentration or addition of inorganic cations changes their shape to a rod-like or cylindrical configuration at a much larger size. The model of a spherical micelle is known to be consistent with the results of small-angle neutron scattering, but this structure is so dynamic that in a small

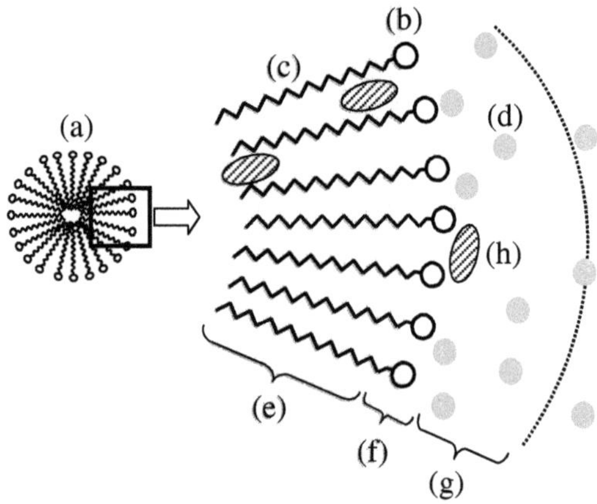

Fig. 3 Schematic representation of spherical micelle: (a) spherical micelle, (b) polar head group of a surfactant, (c) hydrophobic moiety of a surfactant, (d) counterion, (e) hydrophobic core region, (f) Stern layer, (g) Gouy–Chapman double layer, (h) substrate molecule solubilized in micelle

micelle over half the chain segments would be exposed to an aqueous phase at any given time, as demonstrated by nuclear magnetic resonance (NMR) (Bunton and Savelli 1986). A monomeric surfactant rapidly diffuses into micelles and leaves them at much faster rates than thermal chemical reactions; hence, the micelle is considered to be in a pseudo-phase. The cmc values of nonionic and zwitterionic surfactants are generally much less than ionic ones if the size of the hydrophobic part is similar (Table 2). Generally, the longer a hydrophobic alkyl chain, the lower the cmc value. The various physicochemical parameters of surfactant solution and apparent substrate properties change across cmc (Fendler and Fendler 1975; Ying 2006), which can be used to experimentally determine the cmc value. As shown in Fig. 4, the apparent water solubility of a substrate abruptly increases at surfactant concentrations above cmc, while the decreasing surface tension of surfactant solution with an increase of surfactant concentration reaches almost a plateau above cmc.

Table 2 Chemical structure and critical micelle concentration of a typical surfactant

Surfactant	Chemical structure	cmc[a]	AN[b]	T (°C)	Reference
Cationic					
DTMA Br	$C_{10}H_{21}N^{+}(CH_3)_3\ Br^{-}$	1.82×10^{4}	48	25	Fendler and Fendler (1975)
DDTMA Br	$C_{12}H_{25}N^{+}(CH_3)_3\ Br^{-}$	4.62×10^{3}	50	25	
TDTMA Br	$C_{14}H_{29}N^{+}(CH_3)_3\ Br^{-}$	1.18×10^{3}	75	25	
HDTMA Br	$C_{16}H_{33}N^{+}(CH_3)_3\ Br^{-}$	0.33×10^{3}	61	25	
DPC	$C_{12}H_{25}C_5H_5N^{+}\ Cl^{-}$	4.48×10^{3}			Bunton and Savelli (1986)
Anionic					
SDS	$C_{12}H_{25}OSO_3^{-}Na^{+}$	2.3×10^{3}	62	25	Fendler and Fendler (1975)
SHDS	$C_{16}H_{23}OSO_3^{-}Na^{+}$	179		40	
SDSO	$C_{12}H_{25}SO_3^{-}Na^{+}$	2.7×10^{3}	54	25	
SDBS	$C_{12}H_{25}C_6H_4SO_3^{-}Na^{+}$	1×10^{3}			Cho et al. (2002)
Nonionic					
Tween 20	R = monolaurate[c]	60		25	Wan and Lee (1974)
Tween 80	R = monooleate[c]	14		25	
Triton X-45	$(4\text{-}C_8H_{17})C_6H_4O(CH_2CH_2O)_5H$	50	296		Heredia and Bukovac (1992)
Triton X-114	$(4\text{-}C_8H_{17})C_6H_4O(CH_2CH_2O)_{7.5}H$	110	189		Kile and Chiou (1989)
Triton X-100	$(4\text{-}C_8H_{17})C_6H_4O(CH_2CH_2O)_{9.5}H$	130	146		

(continued)

Table 2 (continued)

Surfactant	Chemical structure	cmc[a]	AN[b]	T (°C)	Reference
Igepal CA-720	$(4\text{-}C_9H_{19})C_6H_4O(CH_2CH_2O)_{12}H$	169			Edwards et al. (1991)
Igepal CO-730	$(4\text{-}C_9H_{19})C_6H_4O(CH_2CH_2O)_{15}H$	246	52–80	25	Fendler and Fendler (1975)
Tergitol NP-10	$(4\text{-}C_9H_{19})C_6H_4O(CH_2CH_2O)_{10.5}H$	36.9			Edwards et al. (1991)
Brij-30	$C_{12}H_{25}O(CH_2CH_2O)_4H$	8.3			Krogh et al. (2003)
Brij-35	$C_{12}H_{25}O(CH_2CH_2O)_{23}H$	74			Kile and Chiou (1989)
Neodol 91–6	$C_{9\text{-}11}H_{19\text{-}23}O(Ch_2CH_2CH_2O)_6H$	250	242		Heredia and Bukovac (1992)
Neodol 91–20	$C_{9\text{–}11}H_{19\text{–}23}O(CH_2CH_2O)_{20}H$	390	64		
Silwet L-77	n = 7–8[d]	600			Locke et al. (2002)
Miscellaneous					
Monorhamnolipid	$R = (CH_2)_6CH_3$[e]	54			Mata-Sandoval et al. (2000)
DDAPS	$C_{12}H_{25}(CH_3)_2N^+(CH_2)_3SO_3^-$	402		25	Fendler and Fendler (1975)

[a]Critical micelle concentration in mg/L.
[b]Aggregation number.
[c]Nonionic surfactant (b) in Fig. 2.
[d]Nonionic surfactant (e) in Fig. 2.
[e]Miscellaneous surfacatant (b) in Fig. 2.

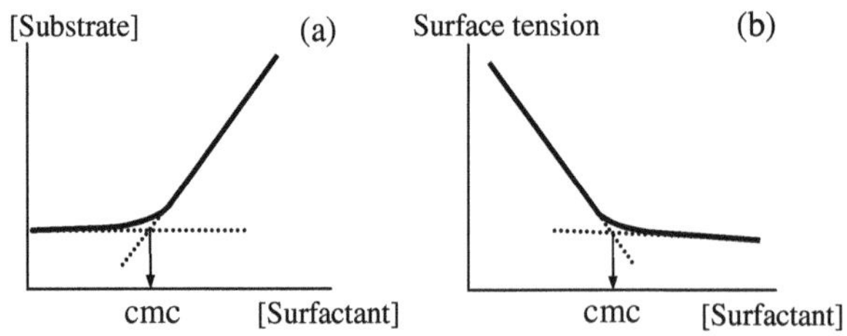

Fig. 4 Effect of surfactant on (a) apparent solubility of a substrate and (b) surface tension of solution. *cmc*, critical micelle concentration

In the simplified model of a spherical micelle, the core region made of alkyl chains is considered to be hydrophobic and fluid and can solubilize an apolar substrate (Bunton et al. 1991; Taşcioğlu 1996). As shown in Fig. 3, the outer region of the micelle consists of polar head groups known as a Stern layer where a part of the counterions are tightly bound in the case of ionic surfactants and the polarity is similar to that of ethanol. The fraction of bound ions is known to be 0.6–0.9 per ionic head group, and the remaining counterions make a diffuse electrical double layer by their thermal motions known as a Gouy–Chapman layer. Polar substrates such as pesticides are then considered to be solubilized in the neighborhood of the Stern layer. In the case of nonionic surfactants, the role of an ionic head group is replaced by the polyethoxy moiety, whose structure may resemble that of a crown ether (Mishra et al. 1992). Raman spectroscopy of octylphenoxy ethoxylates has shown that the helix/coil conformation is predominant for the polyethoxy chain in the presence of excess water via hydration of two water molecules per ether linkage (Bartlett and Cooney 1986). Additionally, the mixing of surfactant with oil such as hexane and long-chain alcohol (cosurfactant) in water gives a spherical microemulsion whose radius is much larger than that of a micelle (Paul and Moulik 2001). Various spectroscopic investigations have shown that the interior of a microemulsion is more hydrophilic than micelles and is considered to produce a different reaction environment from micelles (Almgren et al. 1980; Gregoritch and Thomas 1980; Mackay 1981).

B Biological Effects

Surfactants, especially nonionic ones, are widely utilized to increase the efficacy of pesticides by modifying spray profiles, the penetration through cuticle structures of weeds, crops, and insects to target sites, and translocation, but unfavorable phytotoxicity to crops is also caused by usage depending on surfactant structure and concentration (McWhorter 1985). Some nonionic surfactants are known to inhibit plant growth by affecting root elongation and disrupting the normal membrane permeability and photosynthetic activity. By means of bioassays on germination and growth either in soil and hydroponic media for 17 higher agricultural plants, Günther and Pestemer (1992) reported that most plants are more sensitive to linear alkylbenzene sulfonate (LAS) and 4-nonylphenol, metabolites of nonionic surfactants, than distearyldimethylammonium chloride. Phytotoxicity tests using 14 different cell suspension cultures have shown that tolerance to 4-nonylphenol is correlated with the formation of unextractable residues via metabolic degradation (Bokern and Harms 1997). Either the root uptake of 4-nonylphenol to shoots or the extent of its metabolic degradation was dependent on the plant species (Bokern et al. 1998), indicating that metabolism is one of the important factors to control the phytotoxicity of 4-nonylphenol.

Because surfactants in pesticide formulation as well as their degradation products are considered to move finally to water, their biological effects should be examined not only for crops but also for aquatic species. Generally, cationic surfactants are more toxic to aquatic species including fish, the water flea, and algae than anionic surfactants (Ying 2006). Acute EC_{50} (50% effective concentration) or LC_{50} (50% lethal concentration) and chronic NOEC (no-observed-effect concentration) values for cationic surfactants range from 0.1 to 10 ppm, whereas those for anionic formulations are 1 to 50 ppm. These values are much less than the corresponding cmc values, indicating that aquatic toxicity stems from a surfactant monomer. Nonionic surfactants are mostly less toxic than the ionic and alkylphenoxy ethoxylates are generally less toxic than their degradates, nonyl and octylphenols. The toxicity of nonionic surfactants to both aquatic and terrestrial organisms has been reviewed by Krogh et al. (2003). Their effects on various biological activities are considered to originate from interactions with bioactive macromolecules such as proteins, peptides, amino acids, and membrane phospholipids (Cserháti 1995). Nonionic surfactants are known to bind readily to various proteins and membranes, which alters the protein structure and the permeability of the membrane. Chattopadhyay et al. (2002) have demonstrated by using various sequences in two-dimensional ^{1}H- and ^{13}C-NMR that bovine heart cytochrome *c* oxidase is stabilized by association with lauryl maltoside via interaction of its polar head group with the protein surface. The fluidity of the plasma membrane of corn roots was found to change with the concentration of nonionic surfactant Brij-58 by fluorescence depolarization techniques (Behzadipour et al. 2001). Either the decrease in maximum buildup of a membrane potential or the enhanced decay of the potential in the chromatophores of *Rhodobacter sphaeroides* by linear alcohol ethoxylates showed the nonspecific interactions of these surfactants with the membrane and their effective concentration correlated well with their biological toxicity (Müller et al. 1999).

The effect of surfactants on soil microbiology was studied in a sandy agricultural soil by using polydisperse LAS (Elsgaard et al. 2001). Except for β-glucosidase activity, all soil parameters including basal respiration and the content of phospholipid fatty acids were found to be sensitive to LAS with EC_{10} values from <8 to 22 mg/kg of dry soil weight, which was likely to originate from interactions of LAS with cell membranes.

Distribution of microorganisms in a water–sediment system is known to be affected by the presence of some surfactants. By using sodium dodecyl sulfate (SDS) and its homologues with different alkyl chains, Marchesi et al. (1991) examined the population dynamics of both free-living and attached bacteria by epifluorescence microscopy in a freshwater microcosm. Alkyl sulfates having decyl or dodecyl chains that were adsorbed to sediments and bioavailable to bacteria were found to cause the adhesion of bacteria to the sediment surface. Bacteria are known to be attached on the surface of various solids, which is related to the hydrophobicity of the cell wall being examined by the contact angle of water on cells (Van Loosdrecht et al. 1987). Linear alcohol ethoxylates can alter this cell wall hydrophobicity, depending on the chain length in alkyl and polyethoxy moieties. The longer the polyethoxy moiety, the less adhesion of *Sphingomonas sp.* strain DGB01o onto solid (Brown and Jaffé 2006). When the alkyl chain was elongated, higher adhesion was observed. The alteration of cell wall hydrophobicity was found to be also dependent on bacterial strains from a biodegradation study of octadecane by four *Pseudomonas aeruginosa* strains in the presence of rhamnolipid biosurfactant (Zhang and Miller 1994).

C Environmental Behavior

The pesticide formulation is sprayed on crops in most cases, except for direct soil application using granules, and therefore surfactants in the formulation are considered to first adhere on foliage and shoots of plants followed by penetration into tissues via the cuticle. Surfactants are considered to partly dissipate via sunlight photolysis and metabolism on the plants (See Fig. 1). When the formulation reaches the ground by direct application, or indirectly if not intercepted by leaves or via washoff from precipitation, the surfactants would be taken up by roots, adsorbed to soil, or metabolized by soil microorganisms.

Behavior in Plants

Plant foliage is covered by heterogeneous cuticular waxes that change widely species by species in thickness, composition, and physical structure (Katagi 2004), and the spreading profile of droplets after spraying is highly dependent on the formulation type (Baker et al. 1983). Because plant cuticles consist of waxes and the underlying cutin polymer layer, the penetration profiles of surfactants have been separately examined. Schönherr et al. (1991) examined the partition of octylphenoxy ethoxylates having the different number of an ethoxy unit (N) to the wax-extracted cuticles of citrus leaf and found good correlation between partition coefficient (K) and N; $K = 8634 \cdot \exp(-0.389N)$. The partition of surfactant to the cuticular wax from an aqueous phase ($K_{wax/w}$) was also examined for 15 monodisperse alcohol ethoxylates having different lengths of the alkyl carbon (C) and the ethoxy unit (E) by using barley leaf waxes (Burghardt et al. 1998). The log $K_{wax/w}$ value of each surfactant was found to be linearly dependent on C and E as follows: $\log K_{wax/w} = -2.73 + 0.54C - 0.23E$. The maximum concentration of surfactant in wax at equilibrium (C_{max}) was only dependent on E; $\log C_{max} = 2.04 - 0.14E$. These results show that the hydrophobicity of a surfactant is one of the important factors for its penetration to cuticles; this may be related to the observation that nonionic surfactants, including alcohol ethoxylates, generally penetrate faster into the waxy leaves of rape and pea than the less waxy leaves of bean and beet (Holloway and Silcox 1985). In the penetration of five alcohols and an nonylphenoxy ethoxylate, Tween 20 and SDS to wheat foliage, the molecular weight of the surfactant was not a key factor, but its physical state on foliage, such as gel, liquid, or solid, seemed to control uptake (Anderson and Girling 1983). Nielsen et al. (2005) have used five nonionic surfactants with molecular weights of 290 to 1350 to investigate their penetration into fathen (*Chenopodium album*) and wheat leaves. For both species the total amount taken up for 24 hr exponentially decreased with molecular weight, and the uptake per unit area ($nmol/mm^2$) was found to be proportional to the applied dose per unit area.

To know not only the penetration of surfactant to cuticle but also its translocation, radiolabeled surfactants have been utilized (McWhorter 1985). Part of the treated radioactivity was incorporated into foliage but with insignificant translocation. Microautoradiograms of transverse sections of tobacco leaf treated with ^{14}C-labeled Tween 80 have clearly shown that the surfactant is located only at the upper epidermal cells, indicating its penetration from the plant surface through cuticle to epidermal cells but with no further movement to the mesophyll (Sugimura and Takeno 1985). As compared with foliar uptake, the behavior of surfactant taken up by roots has been much less investigated. The ^{14}C-labeled Soil Penetrant 3685 (linear alcohol ethoxylates) dissolved in a Hoagland solution was taken up by barley roots depending on the surfactant concentration (Valoras et al. 1974). Doucette et al. (2005) utilized crested wheatgrass to study the root uptake and translocation of two ^{14}C-nonylphenoxy ethoxylates under hydroponic conditions. The larger root concentration factor (^{14}C concentration in roots divided by that in solution) was obtained for more hydrophobic surfactant, which was in accordance with the foliar uptake expressed by $K_{wax/w}$. The ^{14}C distribution in leaves, which was unextractable and so caused by metabolites, was found much less than that in roots by a factor of less than 0.01, indicating insignificant translocation.

The metabolism of surfactants has been investigated mainly for the foliar application of monodisperse ^{14}C-labeled alcohol and alkylphenoxy ethoxylates. Successive chromatographic separation of metabolites in barley and rice plants followed by GC-MS has revealed that octylphenoxy ethoxylates undergo oxidation in the alkyl chain and conjugation with glucose at the terminal hydroxyl group to form polar metabolites, as well as release of the ethoxy units to form nonpolar ones (Stolzenberg et al. 1982, 1984). The transformation of nonylphenol as one possible metabolite was examined in wheat cell suspension cultures, and hydroxylation at the alkyl chain followed by sugar conjugation was the dominant metabolic pathway (Bokern et al. 1996). Holloway and Silcox (1985) reported the distribution of polar and nonpolar metabolites from 1-dodecanol octaethoxylate in eight plant species, each of which was tentatively assigned to the tetraethoxylate derivative and sugar conjugates, respectively.

Behavior in Soil

Surfactants directly applied to or reaching soil are first partitioned between the soil matrix and pore water, and some of them are known to modify wetting and water infiltration of soil. Organosilicone surfactants have been reported to reduce water infiltration rates of wet soil (Singh et al. 2002a). To understand the effect on mobility of pesticide in formulation, the adsorption profiles of surfactant should be investigated. Depending on concentration, surfactant molecules would be adsorbed to soil surface as monomer, hemimicelle, or admicelle, or dissolved in pore water as monomer or micelle (Fig. 5) (Edwards et al. 1994). The cation concentration is known also to modify the structure of adsorbed surfactants, and atomic force microscopy has recently shown such structural changes of $HDTMA^+$ aggregates on

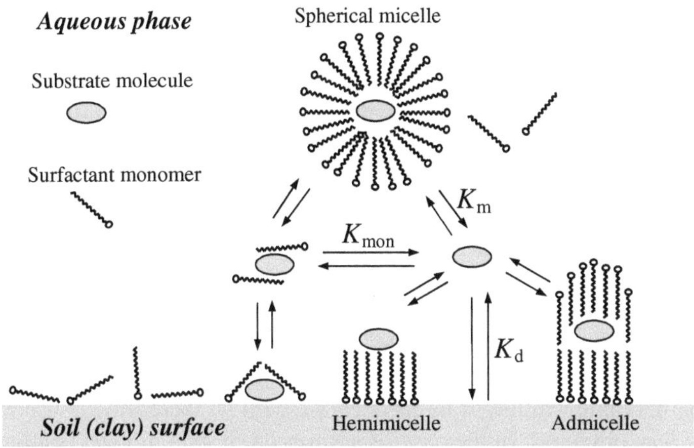

Fig. 5 Adsorption and desorption mechanism of substrate on soil and clay in the presence of surfactant. K_m, micelle-substrate partition constant; K_{mon}, surfactant monomer–substrate association constant; K_d, soil adsorption constant of substrate

muscovite mica (Lamont and Ducker 1998). Soil is a heterogeneous matrix basically consisting of clay minerals and humic substances. The surface of clay minerals is negatively charged by aluminosilicate layers, and humic substances are the heteropolymer of polysubstituted benzenes having various functional groups including hydroxyl and carboxyl groups (Katagi 2004). The interactions between surfactant molecules and soil matrix are then basically classified into electrostatic and hydrophobic. Therefore, both the ionic character of a head group and the hydrophobic chain length of the surfactant as well as soil characteristics, such as organic matter content and clay minerals, are considered to control its adsorption to soil (Haigh 1996; Ying 2006). A multicomponent statistical analysis on adsorption of Triton X-100 was conducted for 21 soil/sediments, and the maximum adsorption was most positively correlated with organic carbon content and clay mineral fraction (Zhu et al. 2003).

Depending on the nature of soil and surfactant, linear, Freundlich and Langmuir isotherms expressed in the following formulae are possible.

$$\text{Linear isotherm: } x / m = K_d \cdot C_w$$

$$\text{Freundlich isotherm: } x / m = K_f \cdot C_w^{1/n}$$

$$\text{Langmuir isotherm: } x / m = (x/m)_{max} \cdot K_L \cdot C_w / (1+K_L \cdot C_w)$$

where x, m, C_w, and n are the quantity of the adsorbed sorbate (surfactant or pesticide), mass of adsorbent (soil), concentration of adsorbate in an aqueous phase at equilibrium, and the constant, respectively. K is the adsorption coefficient with suffixes d, F, and L indicating linear, Freundlich, and Langmuir isotherms, and

$(x/m)_{max}$ is the maximum adsorbate concentration in an adsorbent phase. Irrespective of the isotherm, the presence of soil reduces the aqueous concentration of surfactant because of its adsorption, which results in apparently a larger cmc value in soil–water suspension. In most cases, the Freundlich isotherm is applicable at lower concentrations of surfactant, but the adsorption profiles of nonionic surfactants at a wider concentration range usually obey the Langmuir isotherm, as observed for Triton X-100 and rhamnolipids (Mata-Sandoval et al. 2002).

Liu et al. (1992) examined the adsorption of three nonionic micelle-forming alkylphenoxy ethoxylates and a lamella-forming alcohol ethoxylate below and above their cmc values. Their adsorption followed the Freundlich isotherm below cmc, whereas the maximum plateau values on adsorption of micelle-forming surfactants were obtained above cmc. In the case of the lamella-forming surfactant, the adsorption maximum followed by its decrease with an increasing concentration of surfactant was observed. In the case of adsorption of Triton X-100 on peat soil, a skewed Gaussian-type sorption isotherm was reported, which may be accounted for by the solubilization of soil organic matter by surfactant (Lee et al. 2000). When octylphenoxy and dodecyl ethoxylates were adsorbed on montmorillonite clay, greater adsorption at plateau for the surfactant having a longer ethoxy chain was detected (Lee et al. 2006).

A better correlation of the distribution ratio with the clay fraction in sediment than the organic matter content was observed for adsorption of monodisperse pentadecyl nonaethoxylate (Cano and Dorn 1996). These results show the interaction of a polyethoxy chain with the mineral surface possibly occurs via hydrogen bonding (Krogh et al. 2003). The Freundlich adsorption coefficients of nonionic surfactants are larger than for the anionic ones (Sánchez-Camazano et al. 2003; Urano et al. 1984). A similar trend was reported for adsorption of Triton X-100 and SDS to 18 soils, and the mechanism of adsorption was examined from the aspect of soil characteristics (Rodriguez-Cruz et al. 2005). The K_F values of Triton X-100 were positively correlated with a clay content via interaction with montmorillonite and illite, whereas the largest positive correlation with those of SDS was content of soil organic matter. Therefore, the hydrophobic interaction seems to predominate in the adsorption of SDS and either ion–dipole interaction or hydrogen bonding controls that of Triton X-100. A sediment adsorption study of 10 ^{14}C-labeled LAS surfactants showed that the K_d values increased by a factor of 2.8 for each additional methylene group in the linear alkyl chain, which was in accordance with the hydrophobic mechanism (Hand and Williams 1987). Theoretical assessment using a particle interaction model explained the hydrophobic interactions of several anionic surfactants with soil and sediment, but its contribution is diminished as cation-exchange capacity (CEC) of solid increases (Di Toro et al. 1990).

Greater adsorption of cationic HDTMA Br than SDS was considered to originate from the strong electrostatic interaction of $HDTMA^+$ with the negatively charged sediment surface (Jones-Hughes and Turner 2005). Cationic surfactants tend to be adsorbed strongly onto soil or sediment via an ion-exchange mechanism (Hand et al. 1990). The extent of a surfactant adsorption increased with the length of an alkyl chain in the order of $ODTMA^+ > HDTMA^+ > DDTMA^+$ (Boyd et al.

1988; Lee et al. 1989; Mishael et al. 2002). The exchange profiles of $HDTMA^+$ were more clearly observed for its adsorption to several clay samples, and the higher-charge expandable clay minerals could adsorb more $HDTMA^+$ (Jaynes and Boyd 1991). Depending on the length of an alkyl chain, spacing of silicate layers, and mineral surface charge, alkyltrimethylammonium cations are considered to be incorporated as monolayer, bilayer, pseudo-trimolecular layer, and paraffin complex.

The surfactants dissolved in pore water or surface water and their parts adsorbed on soil and sediment would be degraded by various microbes to more hydrophilic degradates. Existing reviews on microbial degradation of surfactants show more facile degradation under aerobic conditions than anaerobic (Krogh et al. 2003; Ying 2006). LAS is aerobically degraded with half-lives of a few days in river water to about a month in soil and is not degraded anaerobically; alkyl sulfates are readily biodegradable irrespective of areobicity. Cationic surfactants are susceptible to aerobic biodegradation with half-lives about 1 hr but increase of non-methyl substitution at the nitrogen reduces their biodegradability. The nonionic alcohol ethoxylates are degradable in either condition but alkylphenoxy ethoxylates having a longer ethoxy unit become difficult to be metabolized. Each surfactant is generally metabolized by consortia because of the limited metabolic capacity of an individual microorganism (Van Ginkel 1996). Typical metabolic pathways of some surfactants under aerobic conditions are summarized in Fig. 6.

Quaternary trimethylammonium salts are usually degraded by monooxygenase via central fission of the C_{alkyl} — N bond to form trimethyl amine and the corresponding alkanal, which is further oxidized to the acid followed by stepwise β-oxidation. As a minor pathway, ω-oxidation at the far end of the alkyl chain was reported for DTMA Br. Alkyl sulfates are most rapidly biodegradable via enzymatic cleavage of the sulfate ester bond by alkylsulfatases to form the corresponding alcohol and sulfate ion. The major microbial degradation pathway of LAS is the successive ω-oxidation at the far end of the alkyl chain to the corresponding acid, followed by β-oxidation. The final product is a sulfophenyl alkanoate that is further degraded via desulfonation and ring cleavage by dioxygenases. One of the major aerobic routes for alcohol and alkylphenoxy ethoxylates is the oxidation of the terminal ethoxy unit to the carboxylic acid, followed by stepwise β-oxidation finally to form the alkanoic acid, alkylphenol, and alkylphenoxyacetic acid. In the case of alcohol ethoxylates, either central fission to produce the corresponding alkanoic acid and polyethyleneglycol or ω-/β-oxidation at the alkyl chain has been reported. Recently, a unique nitration reaction at the phenoxy ring in soil/sewage sludge mixture has been reported for the possible metabolites of alkylphenoxy ethoxylates (Telscher et al. 2005).

Behavior in Water

Because the chemical class of most surfactants is quaternary ammonium, sulfate, sulfonate, or ether, they are resistant to hydrolysis under environmental conditions.

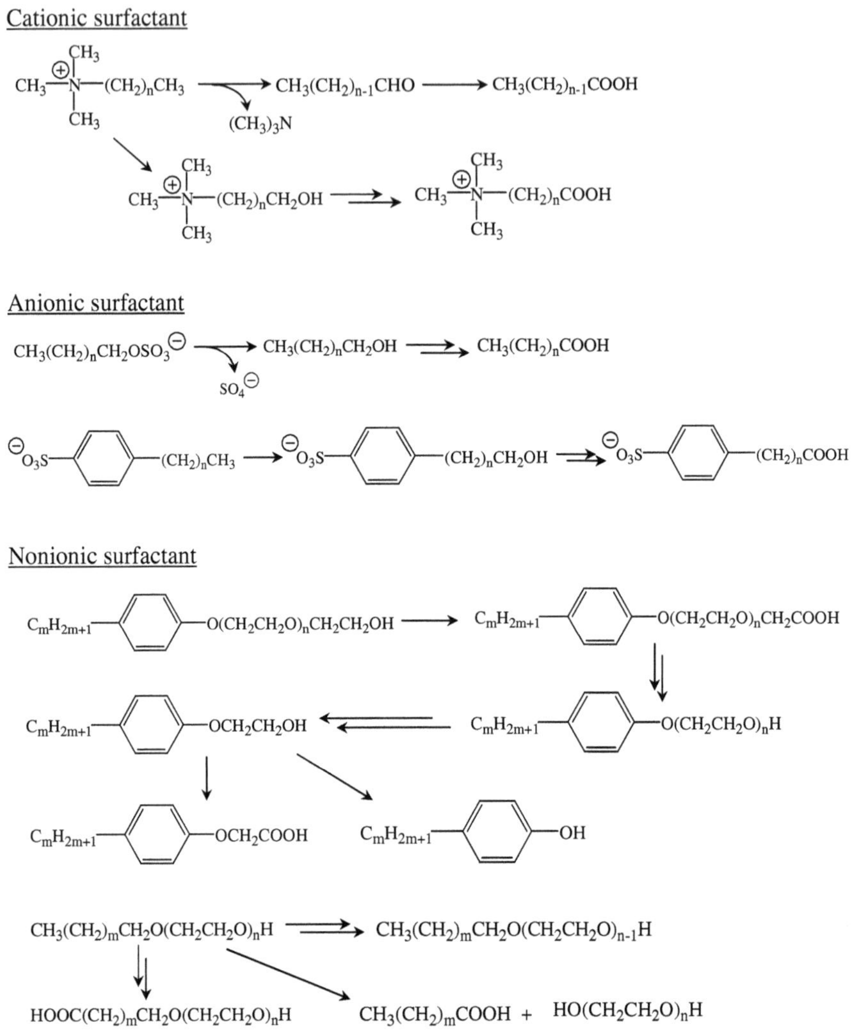

Fig. 6 Primary biodegradation pathways of typical surfactants

Furthermore, many of them are transparent in a UV-visible region from their structures and, hence, direct photolysis initiated by absorbing sunlight is unlikely except for the surfactants having an aromatic moiety such as alkylphenoxy ethoxylates. By using several photosensitizing agents in aqueous photolysis, Tanaka et al. (1991) estimated the triplet energy (E_T) of monodisperse *p*-(1,1,3,3-tetramethylbutyl)phenoxy nonaethoxylate to be 53–54 kcal/mol. This surfactant can act as a photosensitizer for some herbicides but mostly protects photodegradation of herbicides via energy transfer. Photodegradation of the surfactant proceeded via central cleavage of the ether linkage to form the phenol and polyethyleneglycol, the latter of which was further degraded via oxidation or cleavage of a C – C or C – O bond at the terminal ethoxy moiety.

Detailed photoproduct distribution under UV irradiation of nonylphenol ethoxylates has been investigated by liquid chromatography (LC)-MS. (Chen et al. 2007). The branched dodecanol hexaethoxylate was indirectly photolyzed via random ether cleavage with its E_T value of 43–44 kcal/mol (Tanaka et al. 1986).

In a water body, surfactants are partly adsorbed to suspended particles such as clay and biota and also associated with dissolved humic substances. Small-angle X-ray scattering technique has shown the interactions between quaternary ammonium surfactant and soil humic acids that modify the micelle structure (Shang and Rice 2007). Either the association with various suspended and dissolved matters in water or formation of micelles at concentration above cmc is considered to reduce the bioavailability of surfactant. The monomer of surfactant is usually taken up by fish via the gills, but its hydrophilic moiety may make it difficult to pass through the membrane. The concentration ratio of LAS between aqueous and biota phases is widely distributed from 8 to 4100 depending either on species or on alkyl chain length (Tolls et al. 1994). The uptake rate seems to be controlled by hydrophobicity of the surfactant. Alkylphenols metabolized from the corresponding surfactants are also known to be bioaccumulated with factors of 20–1300 (Ying 2006).

III Effects on Physicochemical Properties

A *Solubilization*

A surfactant increases the apparent water solubility of a chemical, and its extent depends not only on the polarity and size of a solubilized chemical but also on the nature of the surfactant such as cmc, aggregation number, and shape of micelle. There are many excellent reviews on enhanced solubilization by surfactants (Cordes and Gitler 1972; Fendler and Fendler 1975; Grieser and Drummond 1988; Taşcioğlu 1996; Thomas 1980; Turro et al. 1980). In general, the apparent water solubility increases in proportion to the concentration of a coexisting surfactant slightly up to cmc and markedly above cmc. The slope in Fig. 4a above cmc is referred to as the molar solubilization ratio (MSR), defined next:

$$\mathrm{MSR} = (S_{mic} - S_{cmc}) / (C_s - \mathrm{cmc})$$

where C_s, S_{mic}, and S_{cmc} are concentration of surfactant (mol/L), and apparent solubilities of substrate (mol/L) at $C_s > \mathrm{cmc}$ and $C_s = \mathrm{cmc}$, respectively (Edwards et al. 1991). When the solubilization is considered as partition of a substrate molecule into micelles, the following equation can also be defined:

$$K_m = X_m / X_a$$

where K_m is a micelle partition coefficient and X_m and X_a are the mole fraction of a substrate in the micellar pseudo-phase and aqueous phase, respectively. When the concentration of surfactant is above cmc, the aqueous concentration as a monomer remains constant at cmc. As X_m and X_a can be expressed by using S_{mic}, S_{cmc}, C_s, and cmc as follows, K_m can be rewritten using the MSR value and the molar volume of water (V_w, 0.01805 L/mol):

$$X_m = (S_m - S_{cmc}) / (C_s - \mathrm{cmc} + S_{mic} - S_{cmc}) = \mathrm{MSR} / (1+\mathrm{MSR})$$

$$X_a = S_{cmc} \cdot V_w, \; K_m = \mathrm{MSR} / (S_{cmc} \cdot V_w)(1+\mathrm{MSR})$$

As another approach, the solubilization can be kinetically analyzed for the equilibrated reaction of $[C] + [D_N] \leftrightarrow [C \cdot D_N]$ with a micelle binding constant of K_s (M^{-1}), where C, D_N, and $C \cdot D_N$ are substrate, micelle, and substrate–micelle complex, respectively. This approach is widely used to kinetically analyze an abiotic hydrolysis in the micellar pseudo-phase model.

The substrate molecule solubilized in micelles is considered to be distributed in accordance with a Poisson statistics with a probability as expressed here (Turro et al. 1980):

$$P(n) = (<s>^n / n!) \cdot \exp(-<s>)$$

P(n) is a probability that the "n" molecule(s) are distributed in a micelle and $<s>$ is a mean occupancy number defined as bulk concentration of a substrate divided by that of micelles. When $<s>$ is 0.1, about 90% of micelles are vacant and the remaining 10% is occupied by a single substrate molecule. The solubilized molecule is considered to spend its time on average with the characteristic region of the micelle, that is, the micellar core (inner and outer), or Stern or Gouy–Chapman layers (see Fig. 3). The core region consisting of alkyl chains is a hydrophobic hydrocarbon-like environment with a higher microviscosity, whereas the Stern layer is rather hydrophilic, where 50%–70% of the ionic head groups is neutralized by counterions in the case of ionic surfactants (Cordes and Gitler 1972; Fendler and Fendler 1975). For nonionic surfactants, Raman spectroscopy, fluorescence polarization of a probe molecule, and the ease of a diffusion-controlled excimer formation, have shown that the long ethoxylated chain of Brij 35 in the Stern layer generally has a dihedral helical structure but the shorter one in Triton X-100 has an open coil structure (Thomas 1980). In the Gouy–Chapman layer, generally with a thickness of several hundred angstroms, an ionic molecule is less tightly associated. Both electrostatic and hydrophobic factors would play a role in determining the solubilization locus of substrate to the micelle, and the various spectroscopic techniques have been applied to estimate the solubilization locus of a substrate. As the ring current of an aromatic substrate perturbs the chemical shift and line width of surfactant protons in the neighborhood of a solubilized site, NMR spectroscopy has been mostly utilized to investigate a solubilized locus (Grieser and Drummond 1988; Taşcioğlu 1996). Because a pesticide is generally a complex polar molecule, its interactions with surfactant would be difficult to interpret. Therefore, the solubilization of simple organic molecules are reviewed first to determine the effect of each functional group or chemical structure on solubilization, and then pesticide examples are discussed.

Simple Organic Molecules

In relation to their bioavailability and soil–water partitioning, solubilization of PAHs by surfactants has been extensively studied, for example, by using naphthalene, phenanthrene, and pyrene. The apparent water solubility of PAH increases with the concentration of nonionic surfactants with their micelle partition coefficients (log K_m) proportional to octanol–water partition coefficient (log *P*) (Edwards et al. 1991; Grimberg et al. 1995). MSR values ranging from 0.04 to 0.3 with log K_m of 4.5–6.5 in the presence of nonionic surfactants showed the high ability of micelles to solubilize PAH molecules. Ionic surfactants also solubilize a variety of hydrophobic organic chemicals including PAH with similar correlation between log K_m and log *P* (Cho et al. 2002; Valsaraj and Thibodeaux 1989). In the case of cationic pyridinium surfactants, decreasing carbon number in the hydrophobic chain reduced the log K_m value (Paria and Yuet 2006). Because the slope of log K_m versus the log *P* plot is less than unity, the micellar phase solubilizing organic chemicals is considered to be less hydrophobic than octanol. The observed correlation with log *P* implies that the most probable site of solubilization for hydrophobic chemicals is a micellar core, but it depends on the size and structure of solubilizate and surfactant molecules. Guha et al. (1998) demonstrated through solubilization of PAH mixtures by Triton X-100 micelles that a less hydrophobic PAH such as naphthalene is located at the micellar core–Stern layer interface, whereas the more hydrophobic phenanthrene is mainly in the micellar core. This difference in the solubilization site as well as competitive solubilization of PAHs having similar hydrophobicity was also confirmed by changes in ^{1}H-NMR chemical shifts of surfactant hydrogens (Bernardez and Ghoshal 2004).

Multiregression analysis on solubilization of hexachlorobenzene in various micelles has shown that both MSR and K_m are dependent on hydrophobicity of surfactant (Jafvert et al. 1994). More carbons in the alkyl or aralkyl chain and fewer ethoxy units resulted in larger MSR and K_m values. Liu et al. (2000) also reported the relationship of log K_m with log *P* and, furthermore, hydrophobicity of a surfactant molecule and the total molecular surface area of a solubilizate were found to be good descriptors for log K_m. Simple hydrocarbons such as hexane and cyclohexane tend to be solubilized in a micellar core. Their higher MSR values of 0.5–4.0 in alcohol ethoxylate micelles, which decrease with an increase of HLB indicate that the volume of a micellar core controls the solubilization (Diallo et al. 1994). In contrast, simple aromatic hydrocarbons and chlorinated alkenes were considered to be located in either the core or the Stern layer of micelle via interaction with the ethoxy units of surfactant. The importance of a core volume was thermodynamically shown for alkyl sulfate surfactant micelles (Abu-Hamdiyyah and Rahman 1987). Similar estimation of their solubilization sites has been reported by using the changes in ^{1}H-NMR chemical shifts of HDTMA Br micelles (Eriksson and Gillberg 1966).

In contrast, polar molecules are considered to be mainly solubilized in the Stern layer or the micelle–water interface. Comparison of an triplet decay in phosphorescence between homogeneous organic solution and micelles has shown that dipolar 4-nitro-4'-methoxystilbene experiences a microenvironment of moderate polarity similar to alcohol and glycerol which is the likely environment of a micelle–water interface (Schanze et al. 1985). The characteristic $n \rightarrow \pi^*$ absorption with a shielding effect in ^{1}H-NMR has been conveniently used to estimate the solubilization site of a benzoyl moiety in micelles. Both the dependency of the $n \rightarrow \pi^*$ wavenumber of the benzoyl moiety on the solvatochromic $E_T(30)$ value describing a medium polarity, and the changes in proton chemical shifts of polar head, methylene, and methyl groups of surfactant, have clearly shown that the aromatic moiety is solubilized from the core to Stern layer with the carbonyl group oriented toward the micellar surface (Fendler et al. 1975). The similar orientation of a molecule in the micellar solubilization by HDTMA Br was reported for phenyl alkanoate derivatives where the long alkyl chain protrudes to the core with the ester moiety located near the micelle surface (Al-Awadi and Williams 1990; Tee and Fedortchenko 1997).

Amides, carbamates, and ureas are considered to behave similarly to esters in solubilization to the Stern layer (Broxton and Chung 1986; Broxton and Sango 1983; Mollett and O'Connor 1976), and this type of association was found to cause the change of a micelle structure with a decrease of cmc via interaction of the surfactant head group (Abu-Hamdiyyah and Rahman 1985). Introduction of a carboxyl group at the *o*-position of phenyl alkanoate derivatives caused more restricted orientation of the phenoxycarbonyl moiety in the surface of HDTMA Br micelles (Broxton et al. 1987). Alcohols and phenols have a very polar hydroxyl group, which would make them locate in the micellar surface. ^{1}H-NMR studies using SDS and HDTMA Br micelles together with other spectroscopy have shown that the molecules having a hydroxyl group are solubilized in the Stern layer of micelles, concomitant with modification on a micelle structure and microviscosity (Bertolotti et al. 1989a; Fendler and Fendler 1975; Mata et al. 2006; Suratkar and Mahapatra 2000; Wolfe and von Bünau 1986). The protection of a hydroxyl group by a methyl resulted in deeper penetration of anisole toward the core but it mainly resided in the neighborhood of α- and β-carbons from the surfactant head group. The ionizable functional groups such as COOH and NH_2 make a solubilizate located in the micelle–water interface. Benzoic acid and aniline respectively solubilized in HDTMA Br and SDS micelles are demonstrated by ^{1}H-NMR so that the ionized functional group is associated with each head group of the surfactant with the aromatic moiety orienting toward the Stern layer (Bunton and Minch 1974; Kim et al. 2001; Manohar et al. 1986).

The accumulated evidence on solubilization of small chemicals having a variety of structure in micelles show that either hydrophobic or electrostatic interactions between a solubilizate and surfactant molecule is one of the most important factors in determining its locus of solubilization in micelles together with the size and shape of the solubilizate and micelles.

Pesticides

Basic investigation on solubilization of pesticides in micelles has been less often conducted, probably because of its practical usage in formulation including various components other than surfactant, lower water solubility, and its complex structure. Solubilization parameters in micelles for the pesticides previously reported are summarized in Table 3. Because DDT (1) is very hydrophobic, with its metabolites persistent in soil, its solubilization by surfactants has been studied extensively. The log K_m values above cmc estimated by the solubility enhancement and surface tension measurement studies are larger than 5, similarly to PAHs. Some association of (1) with surfactant monomers below cmc also enhanced its water solubility but with much smaller log K_m values (Kile and Chiou 1989; Chiou et al. 1991). In contrast, less hydrophobic NAA (2) showed much smaller log K_m values in nonionic surfactants, and its interaction with micelles could be confirmed by a shift of UV absorption to a longer wavelength (Heredia and Bukovac 1992).

The solubilization of organophosphorus pesticides was examined in relation to their micellar-catalyzed hydrolysis and the micelle binding constant K_s tends to increase with the log P value of each pesticide. The usage of cosurfactant such as *n*-pentanol enhanced the solubilization of chlorpyrifos (65) in micelles (Kaushik et al. 2005). Han et al. (2006) estimated that fenitrothion (5) is solubilized in HDTMA Br micelles as its aryl moiety is located from the outer core to the Stern layer of the micelles, with its methoxy groups residing near the head groups of the surfactant. Similar orientation of organophosphorus pesticides has been estimated in lipid bilayers of the several *L*-α-phosphatidylcholines by fluorescence depolarization of probe molecules (Katagi 1995). The MSR values of coumaphos (7), atrazine (13), and trifluralin (15) in Triton X-100 micelles were found to be much smaller than those of PAHs as a result of their larger molecular sizes with less hydrophobicity (Mata-Sandoval et al. 2000). For hydrophobic pesticides, van der Waals force is most likely to account for their interactions with the micellar core, and less hydrophobic ones such as organophosphorus pesticides are likely to be located near the micellar head groups via dipole–dipole interactions and hydrogen bonding.

The more hydrophilic pesticides having amide, carbamate, and imide linkages showed smaller K_s values than (5), which were not so sensitive to the head group of surfactant, and their solubilization in the Stern layer was most likely. The longer the alkyl chain in the hydrophobic part of ionic surfactants having the same head group, the larger K_s value was obtained for carbofuran (9) (Arias et al. 2005) and procymidone (10) (Villedieu et al. 1995). In the case of nonionic alcohol and alkylphenoxy ethoxylates, more hydrophilic surfactants with a higher HLB value solubilized well polar pesticides such as alachlor (53) (Xiarchos and Doulia 2006). Sulfonylurea herbicides with a high water solubility are also partitioned to ODTMA

Table 3 Solubilization of pesticide in micelles

No.	Pesticide/structure	Experimental conditions: surfactant	Solubilization parameters: MSR[1]	log K_m[2]	log K_s[3]	References
1	DDT 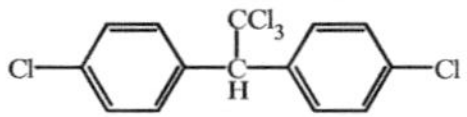	Distilled water at 25±1°C				Kile and Chiou (1989)
		Triton X-100		6.15		
		Brij 35		5.75		
		SDS		5.38		
		HDTMA Br		5.88		
		Distilled water at 25±0.5°C				Cho et al. (2002)
		SDS		5.83		
		SDBS		5.84		
		MADS-12		5.18		
		Distilled water at 24±1°C				Chiou et al. (1991)
		LAS ($C_{10}-C_{14}$)		5.76		
2	NAA 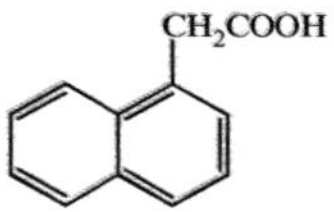	Citrate buffer at pH 3.2 & 23°C				Heredia and Bukovac (1992)
		Triton X-45		3.06		
		Triton X-100		2.66		
		Triton X-405		1.90		
3	Parathion 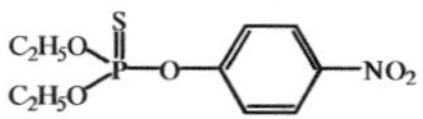	Borate buffer at pH 9.0 & 37°C				Purnanand and Danikhel (1985)
		HDTMA Br			5.69	
4	Paraoxon 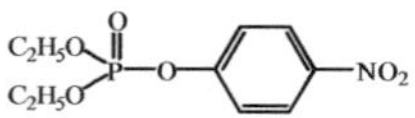	Borate buffer at pH 9.0 & 37°C				Purnanand and Danikhel (1985)
		HDTMA Br			3.88	
5	Fenitrothion 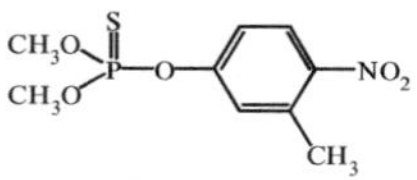	20 mM OH^- at 25°C				Balakrishran et al. (2005)
		HDTMA Br			2.44	
		HDTMA OH			2.40	

6	Cyanofenphos 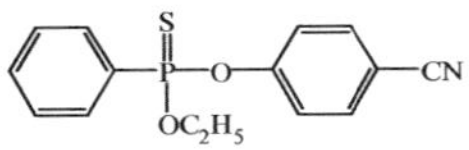	2.5 mM OH^- at pH 11.4 & 30±0.5°C				
		HDTMA Br			3.71	Danikhel and Puranand (1990)
7	Coumaphos 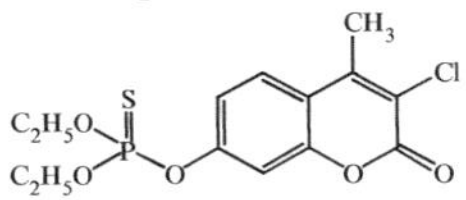	Carbonate buffer at pH8 & 25°C				Mata-Sandoval et al. (2000)
		Rhamnolipid mixtures	2.49×10^{-3}			
		Triton X-100	5.07×10^{-3}			
		Phosphate buffer at pH 7.4 & 25°C				Lopes et al. (1995)
		$C_{12}E_{10}$ [$C_{12}H_{25}O(CH_2CH_2O)_{10}H$]		3.70		
		SDS		4.28		
8	Carbaryl 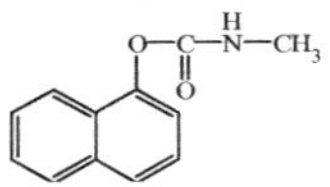	Phosphate buffer at pH 7.5 & 30°C				Patel and Wurster (1991)
		HDTMA Br			2.64	
9	Carbofuran 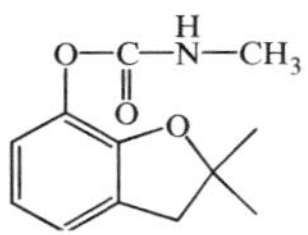	6–7 mM OH^- at 25±0.1°C				Arias et al. (2005)
		SDS			1.80	
		Briji 35			1.46	
		HDTMA Br			1.72	
10	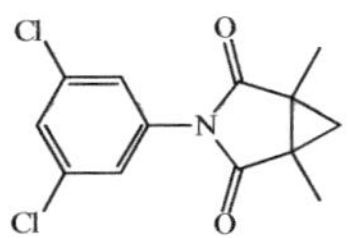	Phosphate buffer at pH 11 & 25°C				Villedieu et al. (1995)
		SDS			2.23	
		HDTMA Br			2.04	
		AA6OE [$C_{13}H_{27}O(CH_2CH_2O)_6H$]			2.07	
11	Sulfometuron	Tetraborate buffer at pH 8.5				Mishael et al. (2002)

(continued)

Table 3 (continued)

No.	Pesticide/structure	Experimental conditions: surfactant	Solubilization parameters: MSR[1]	Solubilization parameters: $\log K_m$[2]	Solubilization parameters: $\log K_s$[3]	References
	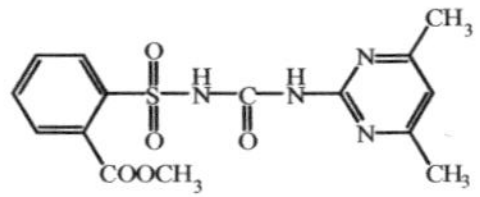	HDTMA Br			3.30	
		ODTMA Br*	0.03–0.10		3.90	
12	Sulfosulfuron	Tris buffer at pH 7.4				Mishael et al. (2003)
	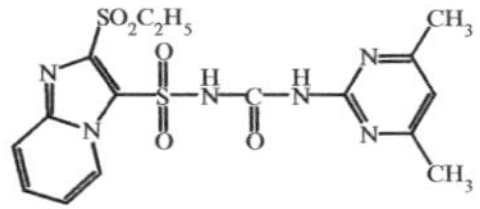	ODTMA Br			3.30	
13	Atrazine	Carbonate buffer at pH8 & 25°C				Mata-Sandoval et al. (2000)
	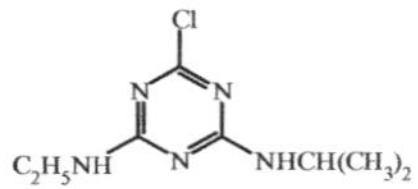	Rhamnolipid mixtures	1.285×10^{-2}			
		Triton X-100	1.07×10^{-2}			
14	Fuberidazole	Phosphate buffer at pH 2.8 & 7.8, 25°C		Neutral / Protonated		Lopes et al. (1992)
	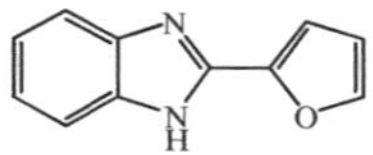	$C_{12}E_{10}$ [$C_{12}H_{25}O(CH_2CH_2O)_{10}H$]		2.28 / 0.58		
		SDS		2.49 / 4.06		
		HDTMA Cl		2.93 / NA		
15	Trifluralin	Carbonate buffer at pH8 & 25°C				Mata-Sandoval et al. (2000)
	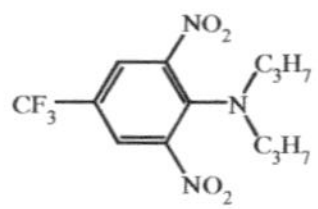	Rhamnolipid mixtures	0.485			
		Triton X-100	9.977×10^{-2}			

LAS, linear alkylbenzenesulfonate; MADS-12, monoalkylated disulfonated diphenyl oxide; ODTMA Br, octadecyltrimethylammonium bromide; Triton X-405, $(4\text{-}C_8H_{17})C_6H_4O(CH_2CH_2O)_{40}H$.

[1]Molar solubilization ratio.

[2]Micelle partition coefficient.

[3]Micelle binding constant (M^{-1}).

Br micelles with log K_s values of 3.3–3.9, and greater than 80% of the applied sulfometuron (11) and sulfosulfuron (12) was incorporated (Mishael et al. 2002, 2003). The effect of electrostatic interactions on solubilization was examined for neutral and protonated fuberidazole (14), and the protonated species was found to be markedly incorporated into anionic micelles (Lopes et al. 1992). In the case of the ionizable 2,4-D (26), electrostatic attraction operates in its adsorption to micellar flocculates of SDS-Al^{3+} (Porras-Rodriguez and Talens-Alesson 1999). Enhanced solubilization has been reported for some pharmaceuticals (Chowdary and Rao 2001; de Castro et al. 1999; Gerakis et al. 1993; Sheng et al. 2006).

B Dissociation

When a chemical is incorporated into micelles, it experiences a different surrounding environment from the bulk water phase, with an influence of electrostatic micellar surface potential, especially for ionic surfactants. This different environment would cause a change of dissociation profiles of an ionizable chemical, which has been investigated by potentiometric and spectrophotometric analyses. Through spectrophotometric analysis of pK_a for dyes in micelles of quaternary alkylammonium surfactants, nonelectrostatic and electrostatic effects were found to be important (Mukerjee and Banerjee 1964; Pourreza and Rastegarzadeh 2005). The former effect by micelles can be evaluated by an effective dielectric constant (ε_{eff}) obtained through comparison of the dissociation constant in a water–organic solvent mixed at various proportions with that in an aqueous micellar system (Drummond et al. 1989). The ε_{eff} values were estimated to be approximately 35 for cationic DDTMA Br, nonionic Brij 35, and *n*-dodecyloctaoxyethylene glycol monoether micelles, which are in good agreement with those estimated by the solvatochromic UV absorption study. In the case of the nonionic surfactant, the ε_{eff} value shows that the polyethoxy moiety is associated with 0.5–1 water molecules. The latter effect is important for the micelles consisting of ionic surfactants, and the observed dissociation constant (pK_a^{obs}) can be described (see below) by using the pK_a value without an electrostatic effect (pK_a^{o}), surface potential (φ), and Boltzmann's constant (k_B) (Drummond et al. 1989; Saha et al. 1994).

$$pKa^{obs} = pKa^{o} - e\varphi/2.303k_BT$$

There have been many investigations on the dissociation process effected by micelles for simple organic acids and amines together with some pharmaceuticals. The typical effects on the dissociation are summarized in Table 4. The pK_a value of benzoic acid decreased by 0.56 in cationic HDTMA Br micelles but increased by 0.4 in the nonionic Tween 80 as compared with that in a bulk water phase (Gerakis et al. 1993). The decrease of pK_a in the cationic micelles means that the equilibrium is shifted to ionization by stabilization of the carboxylic anion by the $(CH_3)_3N^+$-alkyl cation, as demonstrated by ^{1}H-NMR study (Bunton and Minch 1974). As the

Table 4 Effect of surfactants on dissociation constants

Surfactant[a]	Conc.	T (°C)	Chemical	Δ p$K_a^{(obs)b}$	Reference
HDTMA Br	3.7 mM	30	4-Nitrophenol	−0.59	Seguchi (1979)
			3-Methyl-4-nitrophenol	−0.56	
			4-Nonylphenol	−1.61	
	1.2 mM	25	Methyl 4-hydroxybenzoate	−0.29	Loh and Volpe (1992)
	10 mM	25	Benzoic acid	−0.56	Gerakis et al. (1993)
			2-Acetyloxybenzoic acid	−0.21	
DAC	40 mM	30	3-Hydroxybenzoic acid	−1.01	Isoda et al. (1997)
			2-Methylbenzoic acid	−0.91	
SDS	5 w/w%	25	4-Methylaniline	+1.58 ~ 1.91	Drummond et al. (1989)
	9.6 mM	25	Methyl 4-hydroxybenzoate	+0.23	Loh and Volpe (1992)
	10 mM	25	Benzoic acid	−0.19	Gerakis et al. (1993)
			2-Acetyloxybenzoic acid	−0.14	
	20 mM	25	Benzimidazole	+1.30	Saha et al. (1994)
SDBS	3.7 mM	30	4-Nitrophenol	+0.01	Seguchi (1979)
			3-Methyl-4-nitrophenol	+0.06	
			4-Nonylphenol	+2.58	
Tween 80	10 mM	25	Benzoic acid	+0.40	Gerakis et al. (1993)
			2-Acetyloxybenzoic acid	+0.15	
			2-Nitrophenol	+0.39	
	20 mM	25	Benzimidazole	−0.60	Saha et al. (1994)
Brij-35	3.7 mM	30	4-Nitrophenol	+0.01	Seguchi (1979)
			3-Methyl-4-nitrophenol	−0.01	
			4-Nonylphenol	+1.05	
$C_{12}E_8$	5 w/w%	25	4-Methylaniline	−0.43	Drummond et al. (1989)
			4-*tert*-Butylphenol	+1.65	

[a]DAC, dodecylammonium chloride, $C_{12}E_8$, $C_{12}H_{25}O(CH_2CH_2O)_8H$.
[b]Δ p$K_a^{(obs)}$ = p$K_a^{(obs)}$ (in the presence of surfactant) − p$K_a^{(obs)}$ (water).

electrostatic effect is calculated to be approximately −1.7 from the surface potential of the HDTMA Br micelle of about 100 mV (Drummond et al. 1989), the nonelectrostatic effect by HDTMA Br micelle may be much greater than that of Tween 80. Isoda et al. (1997) reported similar electrostatic effects of ionic micelles on pK_a^{obs} by using the ultrasonic relaxation absorption method. Similar trends were observed for simple phenols with variations caused by the substituent effects through inductive and resonance terms, whereas the introduction of a long alkyl chain markedly enhanced the difference of pK_a by 1–2 (Seguchi 1979). Therefore, the hydrophobic effect is considered to be one of the dominant factors controlling the pK_a^{obs} value. The pK_a values of anilines and hetero-aromatics including nitrogen atom(s) in anionic micelles were oppositely increased by the electrostatic effects among the protonated solubilizate and anionic head groups of surfactants (Saha et al. 1994; Sarpal and Dogra 1993). The pK_a values of organic chemicals in Table 4 are those specific to the constant concentration of surfactants and are known to vary with their concentration. Bunton et al. (1980) demonstrated surfactant concentration-dependent change of pK_a based on the pseudo-phase ion-exchange model, the details of which are described in Section IV. The pK_a value consisently changes with the surfactant concentration under a constant concentration of a counterion, and otherwise it shows a maximum change at the characteristic concentration of surfactant controlled by an exchange equilibrium of the counter ion with OH^- or H^+.

Although a pK_a study of a pesticide in micelles is rarely found, similar investigations on pharmaceuticals have been reported. The pK_a values of the carboxyl group in micelles have been potentiometrically examined for salicylic aspirin, naproxen, and iopanoic acid (Gerakis et al. 1993). The cationic surfactants lowered their pK_a values, while they were markedly increased in Tween 80 micelles. There was insignificant effect on the pK_a value by SDS micelles. The acid–base equilibrium of pindolol, chlorodiazepam, and diazepam in SDS micelles was found to shift as their pK_a values increased with the concentration of SDS (de Castro et al. 1999). Also, solubilization of vitamins in micelles has been reported by cyclic voltammetry to affect their redox potentials as well as the corresponding peak currents (Jaiswal et al. 2001).

C Volatilization

Volatilization of chemicals, including pesticides, in the environment is considered to proceed via liquid–air and solid–air interfaces. The former is represented by volatilization of the chemical dissolved in an aqueous phase, and volatilization of a pesticide from its formulation applied to plant foliage or washed off to the soil surface is the typical example for the latter. Water solubility and vapor pressure of a chemical are two of the most important parameters controlling its volatilization from water, which is generally evaluated by the Henry's law constant (H, in atm) as expressed below (Staudinger and Roberts 1996):

$$H = (y_i/x_i)\ P_T$$

where x_i, y_i, and P_T are mole fraction of a chemical in aqueous and air phases and total atmospheric pressure, respectively. Henry's law constant increases by a factor of 1.6 for every 10°C increase under ideal conditions. Higher concentration of a dissolved salt generally increases the H value by the salting-out effect, whereas the association of a chemical with dissolved organic matter, suspended solids, and surfactant micelles is considered to decrease it by increasing the apparent water solubility of a chemical. The addition of soil and algae to water as well as LAS resulted in less volatilization of organochlorine and organophosphorus pesticides from water as a result of their adsorption or solubilization into micelles (Lichtenstein and Schulz 1970). The effect of surfactant on the H value has been examined for small volatile organic chemicals. Anderson (1992) reported that the apparent H values of simple aromatics decreased with increasing concentrations of SDS and nonionic Witconol SN70, with their extent depending on the partition coefficient between micelles and water. Similar profiles were obtained for chlorinated alkanes (Valsaraj and Gupta 1988) and alkenes (Kim et al. 2003; Zhang et al. 2006), where their partition to micelles was the key factor to determine the apparent H value. By using the group contribution method, Smith et al. (1987) proposed the simple approach estimating the H value of small organic molecules. In the foregoing investigations, the H value is usually estimated in the equilibrated system consisting of aqueous and air phases by gravimetrically or chromatographically measuring the concentration of a chemical in each phase.

In assessing the volatilization profiles of a pesticide in the environment, it is useful to know either its volatilization rate or amount from water. Maguire (1991) measured dissipation rates of fenitrothion (5) and deltamethrin (69) in EC formulations from seawater after their application either to its surface or subsurface. Volatilization was a predominant dissipation process in the surface application and (5), having a much smaller H value, dissipated faster. However, the subsurface application resulted in the slower dissipation of (5) than (69), which indicated the importance of other factors such as diffusion via micromonolayer and hydrolysis. Assuming the classical two-film mass-transfer model for volatile chemicals, one can estimate the volatilization rate of a chemical from natural water in the field from the corresponding laboratory data by using the constant ratio of volatilization rate of molecular oxygen between laboratory and field, which is independent of surfactant concentration (Smith et al. 1980). Gavril et al. (2006) developed a convenient method of estimating the volatilization rate of a chemical by using reversed-flow gas chromatography. The evaporation rates of ethanol and 1,1, 1-trichloroethane were found to be reduced by the presence of Triton X-100, but with its extent being significant only when more than two monolayers are formed on the solution surface. Adjuvants other than surfactant such as plant oil were found to also reduce the volatilization loss from water by increasing the pesticide solubility (de Ruiter et al. 2003).

The volatilization loss of a pesticide from a solid surface has been investigated using glass or filter paper as models. Holoman and Seymour (1983) reported linear loss of chlorpyrifos (65) in EC formulation from a glass surface, and the addition of alcohol ethoxylate reduced the volatilization rate. By using the gravimetric method, Sundaram (1987, 1995) extensively analyzed the factors in formulation to control volatilization loss of pesticide and various adjuvants. The higher the viscosity of aqueous formulations, the lower the volatilization rate of pesticide, but with some exceptions. The dissipation of pesticide from the applied filter paper followed the zero-order kinetics in an early stage of volatilization but became first order when a longer period of volatilization was considered (Sundaram 1985; Sundaram and Leung 1986). Garratt and Wilkins (2000) developed a convenient method to assess volatilization of pesticide from formulation by using a glass column filled with formulation-applied glass wool. The volatilized pesticide was collected by a C18 SepPak cartridge and analyzed by GC.

Not only the presence of adjuvant in formulation but also the application method was found to determine the volatilization rate and amount of DDT (1), 2,4-D (26) and dieldrin (62) from a Petri dish (Que Hee and Sutherland 1975). The thin film of the a.i. showed almost complete volatilization at very slow rates, while the corresponding formulations gave smaller amounts of loss at higher rates. Da Silva et al. (2001) reported the reduced volatilization of triadimefon ((*RS*)-1-(4-chlorophenoxy)-3,3-dimethyl-1-(1*H*-1,2,4-triazol-1-yl)butan-2-one) from a glass surface as a wettable powder formulation but with a decreased rate. Volatilization is considered to be highly dependent on the vapor pressure of pesticide but the coevaporation with water may play a significant role, as observed. The isopropylamine salt of glyphosate (*N*-(phosphonomethyl)glycine) was resistant to volatilization from glass slides under moderate wind speed (Leung 1994). The coexistence of long alkyl chain fatty acids was reported to significantly reduce volatilization loss of the fungicide pyrimethanil (*N*-(4,6-dimethylpyrimidin-2-yl)aniline), possibly by salt formation (Green et al. 1998).

Different from a glass surface, volatilization of pesticide from a soil surface would be a more complex process because soil is heterogeneous and adsorbs the pesticide. Chapman and Chapman (1986) have demonstrated the importance of volatilization loss for chlorpyrifos (65) applied as EC formulation to sand by comparing its dissipation between closed and open containers. Under controlled air flow, volatilization rates of thiocarbamate herbicides from six soils were found to be inversely proportional to their adsorption constants to soils with the insignificant effect of EC formulation (Ekler 1988). The effect of adjuvant in EC formulation on volatilization from soil was not consistent among replicates for alachlor (53) (Dailey 2004). In a wind tunnel study using five pesticides in EC or SC formulations, volatilization amounts from silty sand soil seemed to be positively correlated with vapor pressure, and a similar trend was observed for their volatilization from French bean leaves (Rüdel 1997). Either matrix effect or enhanced penetration by adjuvants is known to reduce volatilization loss of pesticide from plant foliage. Abietic acid and its esters as a matrix material was found to significantly improve the volatilization loss of chlorpyrifos (65) from marigold plants, but neither nonionic nor anionic surfactants showed an effect on volatilization (Chen et al. 1994). By analyzing the dissipation profiles in dislodgeable and penetrated residues of azadirachtin (25) in spruce seedlings, Sundaram (1997) speculated that the additives in EC formulation enhanced penetration of the pesticide by dissolving epicuticular waxes and reducing the volatilization.

IV Effects on Hydrolysis

One of the most important degradation processes of a pesticide in water is abiotic hydrolysis (Katagi 2002), and the effect of surfactant on its hydrolytic profiles requires investigation also from the aspect of preparing practically stable formulations.

The concentration of surfactant would be mostly below its cmc in the aquatic environment, and the effect of micelles on hydrolysis of a pesticide is generally of minor importance, except with a limited area of water after a runoff event or a sewage treatment plant. There are many excellent reviews of the effects of micelles on hydrolysis of many kinds of organic chemicals (Bunton and Savelli 1986; Bunton et al. 1991; Cordes and Gitler 1972; Fendler and Fendler 1975; Taşcioğlu 1996).

A Mechanism

Abiotic hydrolysis of pesticides in water mostly obeys first-order kinetics, and the observed hydrolysis rate (k_{obs}) can be expressed by summing the reaction rates of specific acid-catalyzed, neutral, and specific base-catalyzed hydrolysis (Katagi 2002). As hydrolysis is a bimolecular reaction between the pesticide molecule and reactants such as H_2O, H^+, and OH^-, both the solubilization site of pesticide in the micelles and the electrostatic field generated by the ionic head groups on the micellar surface are considered to significantly affect the reaction kinetics. The concentration and medium effects are known for reactions in aqueous micellar systems (Taşcioğlu 1996). The former effect originates from either solubilization of a nonionic reactant in micelles or electrostatic attraction of an ionic reactant to the micellar surface with an opposite charge. In practicice, many pesticides are considered to be located in the Stern layer; the extent of the reactant accessing this region would determine if the micellar reaction is catalyzed or inhibited. The latter effect consists of many factors. The cage effect forces reactants in the neighboring space and enhances a reaction probability. Preorientation effect means the specific orientation of a solubilizate in micelles and may control regio- and stereoselectivity, together with increased microviscosity in micelles. Furthermore, a transition state having a partially ionic character may be stabilized in micelles.

To describe reactions in the micellar system, many kinds of kinetic approaches, including a Hill model for an enzymatic reaction (Piszkiewicz 1977), have been examined. In general, by assuming the two-phase model in Fig. 7, the observed pseudo-first-order rate constant (k_{obs}) can be expressed below (Fendler and Fendler 1975):

$$k_{obs} = \{k_w + k_m \cdot K_s([S] - cmc)\} / \{1+K_s[S] - cmc)\}$$

The rate constants k_w and k_m are the first-order constants in aqueous and micellar phases, respectively. [S] is a total concentration of surfactant, and K_s is a micelle–substrate partition constant. For the reactions in charged micelles whose surface is taken as a selective ion exchanger, the pseudo-phase ion-exchange (PPIE) model has been successfully applied by assuming that a fraction of the surface (β) occupied by the counter ions is constant (Bunton and Savelli 1986; Bunton et al. 1991). Ion-exchange constant (K_N^X) and β are defined below:

$$K_N^X = [N_w][X_m] / [N_m][X_w] \quad \beta = ([N_m] + [X_m]) / ([S] - cmc)$$

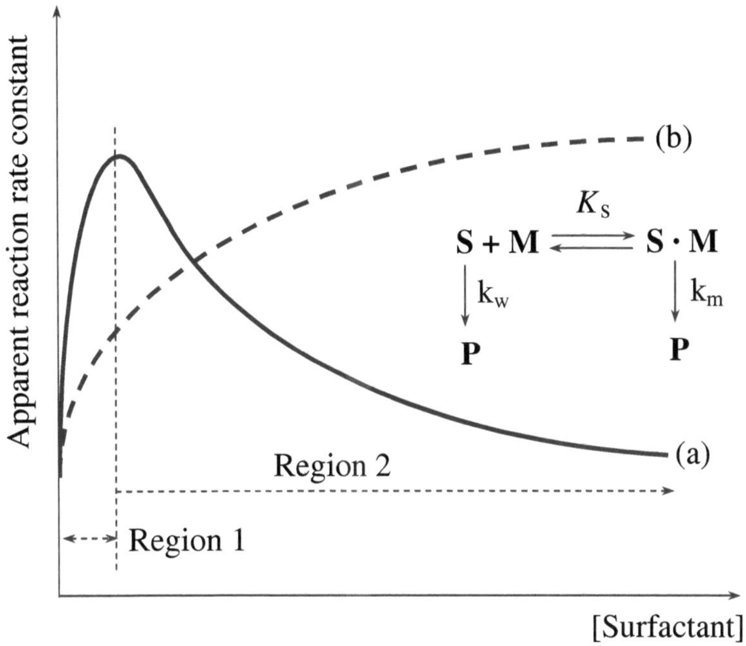

Fig. 7 Base-catalyzed hydrolysis in the presence of cationic surfactant. (a) No adjustment of a counterion concentration. (b) Under the constant concentration of a counterion. *S*, substrate; *M*, micelle; *M.S*, micelle–substrate complex; *P*, product; K_s, micelle–substrate partition constant; k_w, rate constant in an aqueous phase; k_m, rate constant in a micellar phase

N and X are the inert counter ion such as Br^- and the reactive ion such as H^+ and OH^-, respectively. The subscripts w and m represent the bulk water and micellar phases. The β-value is known to be in the range of 0.6–0.9. When the fraction of X in micelles ($m_x = [X_m] / ([S] - cmc)$) is used, k_{obs} can be newly expressed below. [N] and [X] are the total concentrations in the system.

$$k_{obs} = \{k_{2w} \cdot [X_w] + k_{2m} \cdot m_x \cdot K_s([S] - cmc)\} / \{1 + K_s([S] - cmc)\}$$

$$k_w = k_{2w} \cdot [X_w],\ k_m = k_{2m} \cdot m_x$$

$$m_x^2(K_N^X - 1)([S] - cmc) + m_x\{[X] + K_N^X[N] - \beta(K_N^X - 1)([S] - cmc)\} - \beta[X] = 0$$

where k_{2w} and k_{2m} are the second-order reaction rates in the aqueous and micellar phases, respectively. The apparent base-catalyzed reaction rate in cationic micelles generally increases, as shown in Fig. 7 (solid line, region 1), and afterward decreases (region 2), which can be described by the above equation. The k_{2w} and k_{2m} values are mostly of similar order and, therefore, the observed rate enhancement is not a real catalysis but originates from the concentration effect. It is considered that in region 1 the relative concentrations of a solubilizate and X such as OH^- increase in the Stern layer, but the further increase of a surfactant concentration causes competition of available sites in the Stern layer between X and inert counterion N. However, if the concentration of a counterion is kept constant, the lack of further competition between reactive and inert ions results in the continuous increase of the apparent hydrolysis rate (Fig. 7, dotted line). A similar effect of surfactant on hydrolysis in micelles has been reported for oil-in-water microemulsions (Mackay and Hermansky 1981; Mirgorodskaya and Kudryavtseva 2002; Varughese and Broge 1991), and significant change of hydrolysis rate was reported at the phase transition (Hao 2000).

It is known that the PPIE model fails to describe some bimolecular reactions such as nucleophilic aromatic substitution and amide hydrolysis (Broxton 1981; Broxton and Sango 1983). When the reaction center of a solubilizate is located more in the outer side of the Stern layer, reaction with a reactive ion across the interfacial boundary proceeds more readily. Deviation from the PPIE model has been reported for acidic hydrolysis of acetal and aldoxime esters where the apparent rate constants increase in proportion to the acid concentration, which shows the involvement of the reaction with H^+ across the interfacial boundary (Bunton et al. 1979; Gonsalves et al. 1985). A similar deviation in HDTMA Br micelles was also found for alkaline hydrolysis of DDT (1) at high alkaline concentrations (Nome et al. 1982; Stadler et al. 1984).

B Hydrolysis of Pesticides

Simple Organic Chemicals

The structure of the pesticide molecule is so complex in the clear understanding of the effect of micelles on hydrolysis that micellar-catalyzed hydrolysis of simple organic chemicals is discussed first. Kinetic parameters for typical chemicals are summarized in Table 5. The hydrolysis of an organophosphorus ester in micelles has been reviewed extensively by Fendler and Fendler (1975). Its alkaline hydrolysis is affected only by cationic micelles, with the typical profiles in Fig. 7 being observed. Even for the hydrophilic mononitrophenyl esters, the micelle partition constant to HDTMA Br is large (10^4–10^5), and enhanced hydrolysis was found to originate from reduction of the activation energy (Bunton et al. 1968). At the higher pH, the apparent micellar catalysis is more evident for alkaline hydrolysis of *O*-*p*-nitrophenyl *O*,*O*-diphenyl phosphate in HDTMA Br micelles (Bunton et al. 1969). Either anionic SDS or nonionic Igepal micelles were found to markedly inhibit this alkaline hydrolysis, possibly because of the stronger association with micelles and electrostatic repulsion of OH^- from the micellar surface (Bunton and Robinson 1969).

Carboxylic esters are another important chemical class in considering the micellar catalysis of pesticide. Beme et al. (1965) reported the fivefold micellar catalysis in alkaline hydrolysis of phenyl hexanoate in HDTMA Br micelles. Similar to

Table 5 Hydrolysis of functional moieties of organic chemicals in micelles

Functional moiety/structure	Surfactant/medium	T (°C)	K_s[a]	k_{2w}[b]	k_{2m}[c]	Reference
Carboxylic ester						
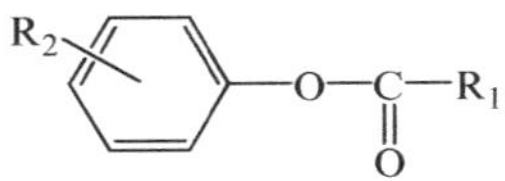						
R_1=CH_3, R_2=H	HDTMA Br (≤30 mM) + 4.7 mM OH^-	30	85	3.2	0.235	Broxton et al. (1988)
	HDTMA SO_4 (<16 mM)		55	3.2	0.182	
R_1=$C_{11}H_{23}$, R_2=H	HDTMA Br (0.1–20 mM) at pH 11.66	25	2.5×10^3	*1.8×10^{-6}	*4.6×10^{-4}	Al-Awadi and Williams (1990)
R_1=$C_{11}H_{23}$, R_2=4-Me	(1–30 mM)		2.4×10^3	*1.0×10^{-6}	*4.0×10^{-4}	
R_1=$C_{11}H_{23}$, R_2=4-Cl	(1–20 mM)		2.6×10^3	*4.6×10^{-6}	*1.3×10^{-3}	
R_1=$C_{11}H_{23}$, R_2=4-NO_2	(1–45 mM)		2.4×10^3	*8.2×10^{-5}	*2.2×10^{-2}	
Amide and anilides						
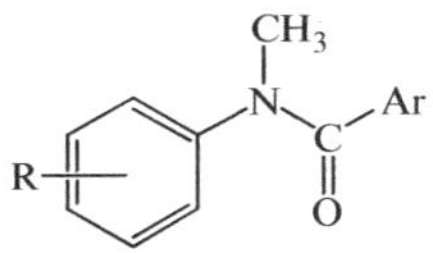						
R=2-NO_2, Ar=Ph	HDTMA Br (<20 mM) + 4.7 mM OH^-	70	200	1.69×10^{-3}	1.4×10^{-3}	Broxton et al. (1988)
	HDTMA SO_4 (<30 mM)		90	1.69×10^{-3}	8.4×10^{-4}	
R= H, Ar=Ph	HDTMA Br (<10 mM) + 5.8 mM OH^-	65.5	172	1.1×10^{-4}	1.3×10^{-5}	Broxton and Duddy (1979)
R= 4-Cl, Ar=Ph			494	2.3×10^{-4}	4.7×10^{-5}	
R= 3-NO_2, Ar=Ph			185	1.37×10^{-3}	9.13×10^{-4}	

(continued)

Table 5 (continued)

Functional moiety/structure	Surfactant/medium	T (°C)	K_s[a]	k_{2w}[b]	k_{2m}[c]	Reference
R= 4-NO_2, Ar=Ph			359	9.07×10^{-3}	3.33×10^{-3}	
R=H, Ar=CH_2Ph	HDTMA Br (<10 mM) + 5.8 mM OH^-	30	841	1.56×10^{-3}	8.8×10^{-4}	Broxton and Duddy (1980)
R=H, Ar=CH_2(4-Me)Ph			2337	1.51×10^{-3}	8.4×10^{-4}	
R=H, Ar=CH_2(4-Cl)Ph			3975	2.26×10^{-3}	1.94×10^{-3}	
R=H, Ar=CH_2(4-NO_2)Ph			348	4.13×10^{-3}	5.47×10^{-3}	
Amide and anilides						
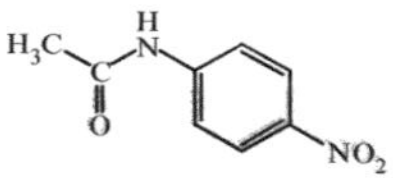	HDTMA Br (15 mM) at pH 9.0	77.85	1.44×10^4	[*]2.68×10^{-5}	[*]1.38×10^{-4}	O'Connor and Tan (1980)
Carbamate						
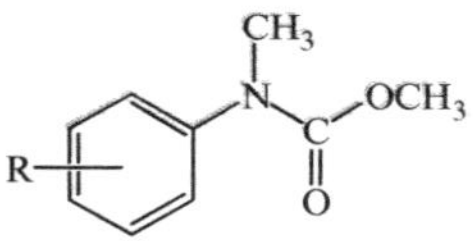						
R=4-NO_2	HDTMA Br (<30 mM) + 9.3 mM OH^-	70	36	1.6×10^{-2}	5.0×10^{-3}	Broxton et al. (1988)
	HDTMA SO_4 (<40 mM)		13	1.6×10^{-2}	4.2×10^{-3}	
R=2,4-$(NO_2)_2$	HDTMA Br (<30 mM)		51	3.6×10^{-2}	3.2×10^{-2}	
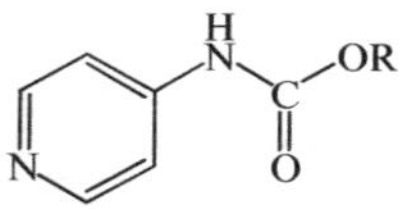	HDTMA SO_4 (<40 mM)		31	3.6×10^{-2}	1.7×10^{-2}	
R=CH_3	HDTMA Br (<6 mM) at pH 13.4	25	46.6		[*]6.35×10^{-6}	Matondo et al. (1990)
	SDS (<1.4 mM) at pH 13.4		120	[*]3.4×10^{-5}		

R=Ph	HDTMA Br (<0.4 mM) at pH 9.18		18.89		*5.02×10^{-3}	
	SDS (<1.4 mM) at pH 13.4		465	*8.0×10^{-4}		
Oxime ether						
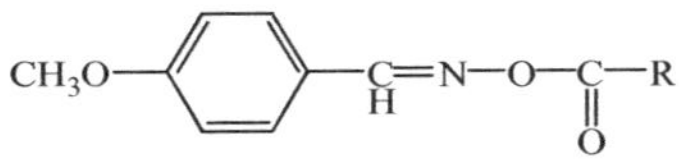						
R=CH_3	SDS (≤ 0.1 M) + 30 mM HCl	25	71	3.11×10^{-3}	6.0×10^{-4}	Gonsalves et al. (1985)
R=C_7H_{15}			3×10^4	3.11×10^{-3}	5.0×10^{-4}	

[a]Micelle binding constant (M^{-1}).
[b]Second-order rate constant in aqueous phase ($M^{-1}s^{-1}$).
[c]Second-order rate constant in micellar phase ($M^{-1}s^{-1}$).
*, First-order rate constant (s^{-1}).

organophosphates, both anionic and nonionic micelles reduced the alkaline hydrolysis rate (Menger and Portnoy 1967). Significant inhibition for the hexanoate in SDS and dodecylphenoxy ethoxylate micelles showed the importance of micellar solubilization of a substrate. The longer the alkyl chain of *p*-nitrophenyl alkanoate, the more enhancement of hydrolysis with the higher micelle partition constant (Tee and Fedortchenko 1997), and the association of the esters with micelles was confirmed by gel chromatography (Romsted and Cordes 1968). By kinetic analysis based on the PPIE model, Broxton et al. (1988) found a lower k_{2m} value for the catalytic alkaline hydrolysis of phenyl acetate in $HDTMA^+$ micelles than k_{2w} by a factor of 14–18, indicating the concentration effect. Similar reduction of hydrolysis rate in the micellar phase was also reported for other phenyl and naphthyl acetates (Vera and Rodenas 1986). Detailed kinetic analysis of alkaline hydrolysis of phenyl laurate derivatives has shown that the catalytic hydrolysis rate obeys the linear Brønsted equation with the β_{lg} value of –0.56, and the effective charge in the transition state is almost the same, irrespective of HDTMA Br micelles (Al-Awadi and Williams 1990). Therefore, the COOPh moiety of the ester was considered to be located in an aqueous-like region of the micelles with its alkyl chain anchoring to the micelle core.

A similar location of the COOPh moiety in the micelles together with the concentration effect has been reported for alkaline hydrolysis of 2-carboxyphenyl alkanoates (Broxton et al. 1987). The introduction of a long alkyl chain at the carbonylcarbon oriented the reaction center as the bulk-phase OH^- was more favorably attacked than that at 4-position of the phenyl ring. The intramolecular general base catalysis is well known for salicylate esters, but solubilization to HDTMA Br micelles has greatly reduced the rate, probably from ion-pair formation between the hydroxyl group of the salicylate and the head group of the surfactant (Kahn and Arifin 1996). In the case of SDS micelles, the alkaline hydrolysis of phenyl salicylate was reduced by its solubilization below the Stern layer, which caused less reaction with water molecules, while that of methyl salicylate was scarcely affected (Kahn 1995).

The effect of micelles on alkaline hydrolysis of amides, anilides, carbamates, and ureas has been studied extensively, and typical kinetic parameters are listed in Table 5. In the case of cationic surfactant, the typical micellar effect explained by the PPIE model was observed, and the hydrophobicity of a solubilizate played a role in the enhanced hydrolysis (Broxton et al. 1978). The reaction mechanism in alkaline hydrolysis such as $B_{AC}2$ and E1cB as well as a rate-determining step is affected differently by micelles. By using *N*-methyl-*N*-phenyl amide and carbamates, the C-N bond-breaking step of the tetrahedral intermediate in the $B_{AC}2$ mechanism has been demonstrated to be affected more in HDTMA Br micelles than the step of OH^- attack on a carbonyl carbon (Broxton 1983, 1984; Broxton et al. 1988). HDTMA Br micelles stabilizes the anionic intermediate of phenyl *N*-(pyridin-4-yl)carbamate in the E1cB mechanism and then enhances alkaline hydrolysis, whereas SDS slows the hydrolysis by hindering the anion formation and repulsion of OH^- from micelles (Matondo et al. 1990).

More detailed analysis on the hydrolysis mechanism of *N*-methyl-*p*-toluanilides in HDTMA Br micelles has been conducted from the standpoint of Hammett correlation (Broxton and Duddy 1979). The curve of the $\log k_{2w}$ versus σ^- plot indicates the change of the reaction mechanism with substitution at the *N*-phenyl moiety, but the constantly higher ρ value of 2.4 in the $\log k_{2m}$ versus σ^- plot indicates that the solvent-assisted C-N bond breaking of the intermediate tetrahedral complex is a rate-determining step in micelles. Similar analysis in HDTMA Br solution has shown that alkaline hydrolysis of *N*-methyl-*N*-phenylacetamides obeys the $B_{AC}2$ mechanism (Broxton and Duddy 1980), and bond forming to a tetrahedral intermediate becomes a rate-determining step for *N,N*-diphenylbenzamides (Broxton et al. 1981). The orientation effect in micellar hydrolysis was clearly demonstrated in alkaline hydrolysis of pentyl *N*-(4-nitrophenyl)carbamate by using differently hydroxylated HDTMA Br surfactants (Broxton and Chung 1986). The replacement of a methyl group by a hydroxyethyl in HDTMA Br showed an insignificant difference in the alkaline hydrolysis rate, but the introduction of OH at the 2-position of the hexadecyl moiety increased the rate by a factor of 1.3–1.7. This change most likely indicates that the carbonyl moiety of the carbamate is located in the Stern layer as the 2-OH oxygen of the surfactant can easily attack. In the hydrolysis of Schiff bases, the entrapment of the protonated bases by the polyethoxylate moiety of nonionic surfactant was considered to control its orientation in micelles (Mishra et al. 1992). Much less investigation of the micellar effect on acid hydrolysis has been conducted. Mollett and O'Connor (1976) investigated this for *N*-phenyl ureas, but neither nonionic nor ionic surfactants significantly affected the hydrolysis rate.

Pesticides

The alkaline hydrolysis of pesticides has been examined mainly in the cationic micelle system (Table 6). The k_{2m} values are generally less than the corresponding k_{2w}, indicating that the apparently enhanced micellar catalysis stems mainly from the concentration effect. In the case of organophosphorus pesticides, the nucleophilic attack of OH^- at the central phosphorus via the S_N2 (P) mechanism is affected by the solubilization site of the pesticide. The more hydrophobic parathion (3) is considered to be favorably located toward the micellar core than its oxon (4), which resulted in the apparent hydrolysis rate of (3) being less dependent on the concentration of HDTMA Br than (4) (Purnanand and Danikhek 1985). The micelle-catalyzed hydrolysis of (4) was inhibited by addition of inorganic salts in the order of NO_3^-, Br^-, and Cl^-, which was consistent with the efficiency of counterions in competitive binding with the micellar head groups. Balakrishnan et al. (2004, 2005) and Han et al. (2006) investigated the hydrolysis of fenitrothion (5) in the $HDTMA^+$ micelles by using various α-nucleophilic oximates. The accelerated hydrolysis was considered to originate from both solubilization of (5) in the Stern layer and attraction of oximate anions to the micelle surface. The hydrophobic oximate such as $Ph(CH_3)C{=}NOH$ is considered to be solubilized in micelles as the hydrophobic $Ph(CH_3)C{=}$ moiety is

Table 6 Hydrolysis of pesticides in micelles

No. Pesticide/structure	Surfactant/medium	T (°C)	k_{φ}[a]	k_{2w}[b]	k_{2m}[c]	Reference
3 Parathion 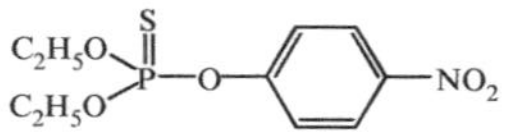	HDTMA Br (≤90 mM) in borate buffer at pH 9.0, ± 0.1–1 M H_2O_2	37	0.9–1.4 × 10^{-4}	*0.41 × 10^{-4}		Purnanand and Danikhel (1985)
4 Paraoxon 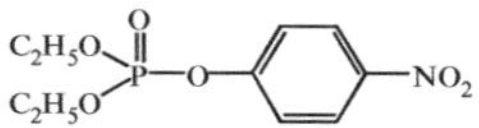	HDTMA Br (≤90 mM) in borate buffer at pH 9.0, ± 0.1 – 1 M H_2O_2	37	3.9–6.6 × 10^{-4}	*2.45 × 10^{-4}		Purnanand and Danikhel (1985)
5 Fenitrothion 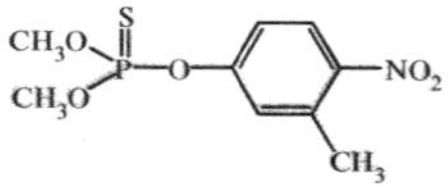	HDTMA OH (20 mM)	25	9.6 × 10^{-4}	2.43 × 10^{-3}	1.66 × 10^{-3}	Balakrishnan et al. (2004)
	HDTMA MINA[d] (20 mM)		8.2 × 10^{-4}	1.15 × 10^{-3}	7.80 × 10^{-4}	Balakrishnan et al. (2005)
	HDTMA Br (≤10 mM) + 20 mM OH^-			2.43 × 10^{-3}	2.89 × 10^{-3}	
6 Cyanofenphos 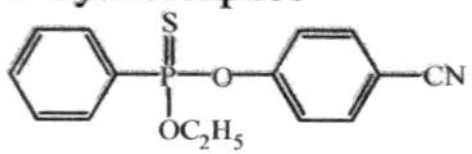	HDTMA Br (≤130 mM) at pH 10 - 11.4, ± 10 – 70 μM H_2O_2	30±0.5	1 – 38 × 10^{-4}	0.10 – 0.16	1.61 – 3.06 × 10^{-4}	Danikhel and Purnanand (1990)
8 Carbaryl 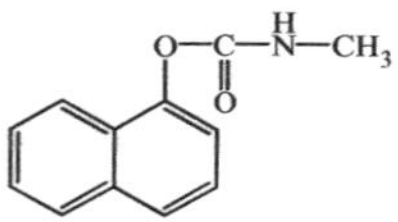	HDTMA Br (≤12 mM) in phosphate buffer at pH 7.5	30		*4.58 × 10^{-6}	*2.05 × 10^{-4}	Patel and Wurster (1991)
	HDTMA Br (5 mM) in 5 natural waters + 3 – 10 μM OH^-	25±0.1	6 – 22 × 10^{-4}	2–11 × 10^{-4}		Gonzalez et al. (1992)
9 Carbofuran 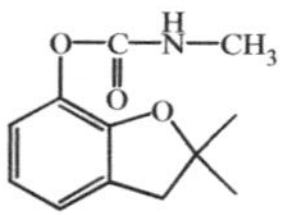	SDS (≤0.12 M) + 6 – 7 mM OH^-	25±0.1	1–8 × 10^{-3}	1.16		Arias et al. (2005)
	HDTMA Br (≤0.15 M) + 1 mM OH^-		0.01 – 0.27	1.16	1.8±0.3	
	Brij-35, Brij-58, Brij-78[d] (≤0.06 M) + 6 mM OH^-		2.8 – 7 × 10^{-3}	1.16		

10 Procymidone 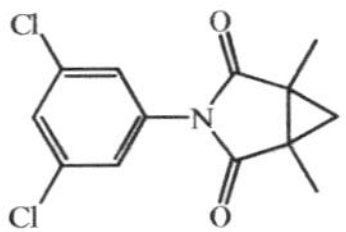	HDTMA Br (≤5 mM) in phosphate buffer at pH 11	25		$*5.62\times10^{-3}$	$*4.4\times10^{-4}$	Villedieu et al. (1995)
16 Iprodione 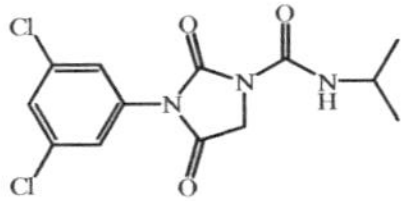	HDTMA Br (≤5 mM) in borate buffer at pH 9.1	25		$*1.15\times10^{-3}$	$*6\times10^{-5}$	Villedieu et al. (1995)
17 Vinclozolin 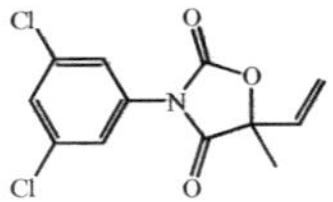	HDTMA Br (≤5 mM) in phosphate buffer at pH 10.5	25		$*8.2\times10^{-3}$	$*3.15\times10^{-3}$	Villedieu et al. (1995)
18 Chlozolinate 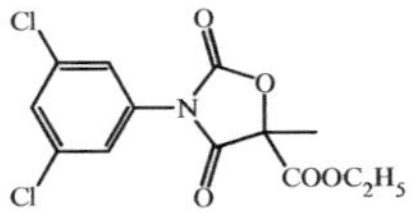	HDTMA Br (≤5 mM) in phosphate buffer at pH 7.1–11.0	25		$*1.8\times10^{-2}$	$*4.3\times10^{-3}$	Villedieu et al. (1995)
19 Amitraz 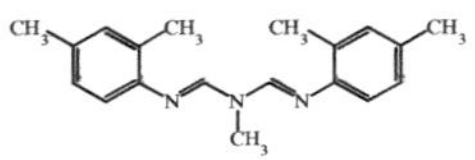	SDS (0.5–2%) HDTMA Br (0.5–2%) Tween 80 (0.0125–0.05%)	25	$1.1–1.3\times10^{-4}$ $1.0–1.3\times10^{-6}$ $4.2–6.1\times10^{-6}$			van Eden et al. (2004)
1 DDT 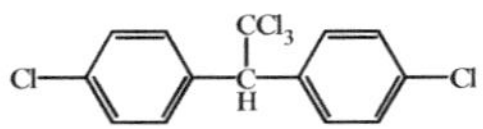	HDTMA Br (1–100 mM) + 0.1–63 mM OH^-	25±0.1			$1.3–1.4\times10^{-3}$	Nome et al. (1982)
	HDTMA OH (1–30 mM)	25±0.1			3.42×10^{-3}	Stadler et al. (1984)

(continued)

Table 6 (continued)

No. Pesticide/structure	Surfactant/medium	T (°C)	k_φ[a]	k_{2w}[b]	k_{2m}[c]	Reference
20 Dicofol 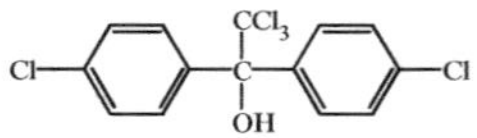	HDTMA Br (≤0.1 M) at pH8.5	30		$*1.8\times10^{-4}$	$*2.4\times10^{-4}$	Nome et al. (1980)
	SDS (1–100 mM) at pH 11	30	0.07–224×10^{-4}	$*2.89\times10^{-2}$		
	Dodecylcarnitine chloride (1–18 mM) at pH 10.3	30	4.0–4.9×10^{-3}	4.18×10^{-3}		

[a]Apparent pseudo-first-order rate constant (s^{-1}).
[b]Second-order rate constant in aqueous phase ($M^{-1}s^{-1}$).
[c]Second-order rate constant in micellar phase ($M^{-1}s^{-1}$).
[d]MINA, *anti*-pyruvaldehyde 1-oximate anion; Brij-58, $C_{16}H_{33}(OCH_2CH_2)_{20}OH$; Brij-78, $C_{18}H_{37}(OCH_2CH_2)_{20}OH$.
*, First-order rate constant (s^{-1}).

located deeply from the Stern layer. By product analysis using ^{31}P-NMR, this oximate was found to hydrolyze (5) to the corresponding *o*-demethylated derivative via the SN2 (C) mechanism in addition to the S_N2 (P) reaction. Therefore, (5) is most likely to be solubilized in the HDTMA$^+$ micelles as its phenyl ring is located in the Stern layer with the P=S moiety protruding toward the micellar surface.

The alkaline hydrolysis rate of carbaryl (8) exhibited the typical dependency of the HDTMA Br concentration (see Fig. 7a) (Gonzalez et al. 1992). From the temperature dependency of the hydrolysis rate, the activation enthalpy was estimated to be 16.4 cal/mol·K, slightly larger than that without micelles, and therefore the reaction was considered to obey the E1cB mechanism where the formed nitranion was stabilized by association with the positively charged head groups (Patel and Wurster 1991). The enhanced alkaline hydrolysis was reported for carbofuran (9) in the cationic micelles, while at higher concentrations of alkyl sulfates and alkoxy ethoxylates it was markedly retarded (Arias et al. 2005). The longer the alkyl chain of the cationic surfactants, the more enhancement of hydrolysis was observed in the micelle system. When the maximum hydrolysis rate (k_{max}) in the micelle system was used, the $\log(k_{max}/k_{2w})$ versus $\log K_s$ plot was linear, indicating the importance of solubilization of (9) in the micelles. The retardation of alkaline hydrolysis in the anionic micelles can be accounted for by an ionic repulsion of OH^- from the micelle surface, but the inhibitory mechanism by the nonionic micelles was not clear. Similar hydrolytic profiles have been reported for dicarboximide pesticides such as procymidone (10) (Villedieu et al. 1995). The reduced hydrolysis rates in nonionic micelles were explained by the significant association between the pesticides and micelles, and the polyethoxy chain might make the nucleophile OH^- less accessible to the carbonyl carbon of the pesticides. Similar interactions would play a role in more favorable decarboxylation in the side chain of chlozolinate (18) after ester cleavage in the nonionic micelles, and stabilization of the transition state not involving water molecules was postulated. Similar hydrolytic resistance for captan (47) and phosmet (66) having the dicarboximide moiety was reported for the corresponding wettable powder formulations (Atwood et al. 1987).

Dehydrochlorination of DDT (1) under the alkaline conditions has been examined in the HDTMA Br micelles at a wide range of OH^- concentration (Nome et al. 1982). At the lower OH^- concentration of 10^{-3}–10^{-2} M, the micellar catalysis was well described by the PPIE model, whereas additional reaction took place across the interfacial boundary between the Stern and Gouy–Chapman layers of the micelles at the higher OH^- concentration (Stadler et al. 1984). Deviation from the PPIE model is considered to originate from the very dynamic surface structure of the micelles. Further, addition of a long-chain aliphatic alcohol such as hexanol was found to reduce micellar catalysis in the HDTMA$^+$ micelles from that expected from the PPIE model (Otero and Rodenas 1986). The incorporation of the long-chain alcohol increased the volume of micelles, and the decrease of an effective concentration would result in a reduced hydrolysis rate. Micellar catalysis by cationic surfactants was observed for dicofol (20) mainly via the concentration effect, which was confirmed by the positive value of an activation entropy (Nome et al. 1980). Inhibition of hydrolysis in the SDS micelles together with an insignificant effect by zwitterionic dodecylcarnitine chloride surfactant was also reported. The effect of micelles on acidic hydrolysis of pesticide has been less investigated. Van Eden et al. (2004) examined the hydrolysis of amitraz (19) in SDS, HDTMA Br, and Tween 80 micelles. The hydrolysis rates in the latter two micelles were found to be less than that in water, while SDS micelles enhanced hydrolysis by a factor of 5–10. The observed enhancement of acid hydrolysis can be accounted for by an attracting proton on the surface of micelle where (19) is solubilized. A similar profile was expected for the acid-catalyzed hydrolysis of atrazine (13) in the anionic SDBS micelles, but its effect was found to be minimal (Lei et al. 2001).

V Effects on Photolysis

Photolysis is considered to be another important route of pesticide degradation in formulation. During some period of water evaporation from the applied formulation on the surface of foliage and soil, adjuvants in the formulation may affect the photodegradation profiles of pesticide. Furthermore, because sunlight irradiance is known to change only slightly in shallow water (Barron et al. 2000) and surfactants in formulation tend to gather in the surface microlayer of the water (Katagi 2006), the effect of the surfactant on aqueous photolysis may become important in some cases. There are many excellent reviews on the effect of surfactant and micelle on photophysical and photochemical processes (Fendler and Fendler 1975; Ramamurthy 1986; Taşcioğlu 1996; Thomas 1977, 1980; Turro et al. 1980).

A Mechanism

The basic investigation on the effect of surfactants and micelles has focused mostly on photolysis in the solution phase (Fig. 8). The excited singlet or triplet state of a pesticide (S^*) by absorption of light undergoes either chemical reactions or deactivation to the ground state (S) via internal conversion or quenching, emission of fluorescence, or phosphorescence via intersystem crossing and energy transfer. Each process can be significantly affected by not only the polarity and viscosity of the medium where a pesticide molecule is solubilized but also the locus and orientation of the molecule in the micelle. The fine structure of pyrene fluorescence sensitive to solvent polarity showed the hydrophilicity of the solubilized locus in the HDTMA Br micelle, and its enhanced excimer formation implied the higher microviscosity in the interior of the micelle (Thomas 1977, 1980; Turro et al. 1980). The quenching of the excited state in a micelle was markedly controlled by the partition of nonionic quencher (Q) into the micelle as well as the charge of a quencher (Q^+, Q^-) that determined its accessibility to the ionic micellar surface. For a chemical solubilized in the Stern layer, the heavy atom effect by a counterion such as Br^- in HDTMA Br results in a decrease of its fluorescence lifetime by a heavy atom effect

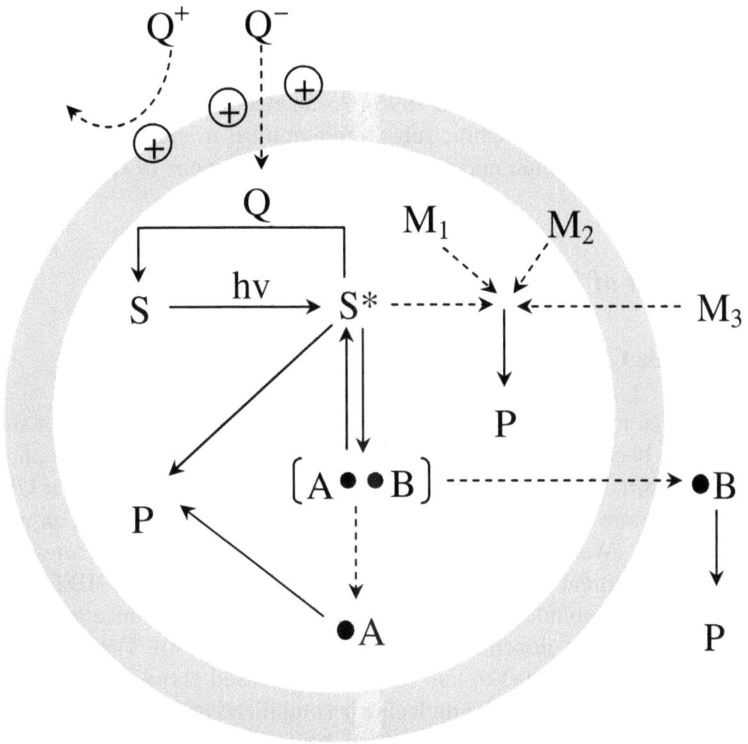

Fig. 8 Nature of photoreactions in micellar systems. *S*, pesticide; *P*, photoproduct; [*A*• •*B*], paired radicals in a solvent cage; *A*• and *B*•, free radicals; M_1, reactant solubilized in micelle; M_2, weakly micelle-associated reactant; M_3, reactant in an aqueous phase; *Q*, quencher solubilized in micelle; Q^+ and Q^-, ionic quenchers. *Dashed* and *solid arrows*, diffusion and reaction processes, respectively

(Fendler and Fendler 1975). When the energy donor and acceptor coexist in the same micelle, the efficiency of energy transfer would increase.

Based on the Poisson distribution, most micelles are considered to be free from molecular oxygen under the usual conditions of $[O_2] = 10^{-4}$ M and [micelle] = 1 mM. The characteristics of a micelle result in several effects on photoreactions, the most popular one being a cage effect (Ramamurthy 1986). The reactive species such as radicals (A· and B· in Fig. 8) produced by photoinduced bond cleavage of a solute molecule spends more time in the restricted micellar space than in homogeneous solution, which alters the profiles of radical recombination. The product distribution in the Norrish type I and II reactions of benzoin methyl ether significantly varied with the type of surfactant-forming micelles. The orientational effect also controls the product distribution in various photoreactions (Turro et al. 1980). The position of intermolecular hydrogen abstraction from the alkyl chain of surfactant by the excited benzophenone derivatives was found to be dependent on the locus of the carbonyl group in micelles. The regioselectivity in photo-induced cycloaddition or dimerization of alkenyl species such as cyclopentenone was found to be highly dependent on the orientation of the solubilized molecule in a micelle with the promoted reaction because of the increase of local concentrations of reactants (Ramamurthy 1986; Taşcioğlu 1996). As an electrostatic effect, the enhancement of photo induced nucleophilic substitution of nitro aromatics by a cyanide ion was reported for the cationic micelles (Hautala and Letsinger 1971).

B Photolysis of Pesticides

Simple Organic Chemicals

Photoinduced isomerization is one of the basic photoreactions, and that of stilbene derivatives has been examined extensively in the various aggregates. The photostationary isomerism of 4-hydroxystilbene in the HDTMA Br micelles under UV irradiation shifted more to the *cis*-isomer than that in methanol because of an increase of microviscosity (Wolfe and von Bünau 1986). No formation of the corresponding dimer and phenanthrene derivative may imply its association with HDTMA Br in the Stern layer. Photoinduced regioselective dimerization in micelles has been reported for 3-*n*-butylcyclopentenone (Lee and de Mayo 1980). The isomeric ratio of dimers, head-to-head/head-to-tail, markedly increased above cmc, indicating a preorientational effect so that the molecule is solubilized in the Stern layer with its butyl chain inserted toward the micelle core. A similar effect on regioselectivity in dimer formation in micelles was reported for uracil (Fendler and Bogan 1974) and cinnamic acid (Nakamura 1988). Photoinduced polyene cyclization is known to proceed more favorably in micelles, presumably as a result of stabilization of intermediate radicals (Hoffmann et al. 1993).

Photoinduced dechlorination is a common reaction in pesticide photochemistry. Photodegradation of trichloroethane in the Brij 35 micelles at 253.7 nm was enhanced by a factor of 3 more than that in water (Choy and Chu 2001). Much higher quantum yield of photodechlorination via electron transfer mechanism in ionic micelles than water was reported for many polychlorinated benzenes (Chu and Jafvert 1994). The lack of formation of biphenyls shows the compartmentalization effect of the micelle that results from isolation of the aryl radical from other aryl molecules. This technique was applied to efficiently decompose polychlorobiphenyls in soil (Chu et al. 1998). Product analysis showed that the dechlorination proceeded stepwise by using the surfactant as a hydrogen source finally to the corresponding nonchlorine aromatics or phenol (Chu and Jafvert 1994; Shi et al. 2000). The photoinduced dechlorination of 2-chlorphenol to form phenol proceeded in higher quantum yields with the additional products when dissolved in micelles than that in water (Shi et al. 1997). The nucleophilic displacement of Cl in the undissociated species with water under UV irradiation gave catechol, while the anionic species underwent Wolff rearrangement to form cyclopentadienecarboxylic acid. The photoreduction of nitroaromatics in the alkaline borohydride solution was promoted by the cationic micelles via association of the reactive BH_4^- species on the micellar surface but was inhibited by anionic micelles due to electrostatic repulsion (Larson et al. 2000). In addition to the photoreduction as above, the effect of micelles on photooxidation by $^1O_2$ was also examined. The oxidation of polychlorophenolate ion by $^1O_2$ was enhanced in the HDTMA Cl micelles (Bertolotti et al. 1989b). In contrast, the oxidation of nitrophenolate ions by $^1O_2$ was retarded in the HDTMA Br micelles although both reactive species were solubilized near the micellar surface (Luiz et al. 1992). Since oxidation proceeds via electron transfer from the nitrophenols, association with the cationic micelles is considered to inhibit this process. The reduced photooxidation of pyrene in the Brij 35 micelles was examined at 350 nm by Sigman et al. (1998). The reaction was likely to be initiated by electron transfer from pyrene instead of its excited triplet as there was no inhibition of the reaction by Hg^{2+}. This process seemed to be affected by the association of pyrene with the polyethoxy moiety of Brij 35 estimated from the resemblance of the fine structure of fluorescence in triethylene glycol.

The Norrish type II reaction of valerophenone was examined in benzene, *t*-BuOH, and HDTMA Br micelles (Turro et al. 1977). The regioselectivity and quantum efficiency in intramolecular cycloaddition via abstraction of the γ-proton by the excited carbonyl oxygen was very similar in the micelles and *t*-BuOH with no quenching of acetophenone formation by Eu^{3+}, showing the solubilization of the ketone in the Stern layer. The spin state of the ketyl radical produced from the excited benzophenone in ionic micelles has been examined under a moderate magnetic field (Scaiano et al. 1982). The reduced intersystem crossing of the radical pairs caused by Zeeman splitting of the triplet state caused less formation of the *para*-coupling product, indicating the singlet pathway. The photoinduced decarbonylation via a free radical mechanism has been studied extensively in relation to the cage effect of micelle by using dibenzyl ketone derivatives (Turro and Cherry 1978; Turro et al. 1987). When A-C(=O)-B (A, phenyl; B, tolyl) was photolyzed at lower concentrations of HDTMA Br around its cmc, the distribution of 1,2-diarylethanes followed the statistics of AA:AB:BB = 1:2:1. In contrast, at the higher surfactant concentrations well above cmc with the small ketone/micelle ratio, the product ratio became <1:98:<1, clearly indicating the cage effect. The similarly high production of AB in the micelle system containing Cu^{2+} implied the formation of AA and BB in an aqueous phase from the released radicals. By using the steady-state chemically induced dynamic nuclear polarization (CIDNP) technique, Turro et al. demonstrated that the escape of the radical (A· or B·) from the secondary geminate radical pair (A· – ·B) produced via decarbonylation from micelle controls the extent of recombination to form AB, and therefore the micelle having a larger volume showed more significant cage effect. Lei et al. (1991) have shown a higher cage effect in HDTMA Br and SDBS micelles than nonionic Triton X-100 one and concluded that the ionic micelles reduce the radical escape by the counterions in the Stern layer. In the case of benzyl esters having one of the basic structures of pesticide such as pyrethroid, a similar cage effect was reported for its photoinduced decarboxylation (Turro et al. 1980; Anvir et al. 1981). The photo-Fries rearrangement of esters and amides was also studied in the micelle system from the aspect of organic synthesis (Singh and Sonar 1985; Singh and Raghuraman 1985, 1986). Both the cage and preorientational effects by SDS micelles resulted in the regioselective formation of *o*-migration products with higher yields than organic solvents.

Pesticides

The effect of adjuvants on photolysis of pesticides has been studied as a deposit or thin film on a glass surface, in aqueous solution, and on soil surface (Table 7). An insignificant effect was observed in many cases, but some adjuvants seem to accelerate photodegradation of pesticides. On the glass surface, Tween-type surfactants significantly promoted photodegradation of chinomethionat (21) by UV light, presumably because of sensitization (Nutahara and Murai 1984). In contrast, these surfactants showed an insignificant effect on photodegradation of sulfonylurea herbicides (Thomas and Harrison 1990). The photostability of azadirachtin-A (25) was investigated by using nine nonionic surfactants, and some of them were considered to promote photolysis via energy transfer to the pesticide (Johnson and Dureja 2002). Enhanced photodegradation of silcotrione (24) in formulation was reported on the film of carnauba wax by using a solar simulator (Halle et al. 2000). A threefold-higher rate of intramolecular cyclization of (24) to the corresponding chromone derivative was observed as compared with the a.i.

Hautala (1978) investigated the effect of HDTMA Br and SDS on photolysis of the methyl ester of parathion (3), carbaryl (8), and 2,4-D (26) on soil thin layers. The surfactants gave insignificant or inconsistent effects on photodegradation of (3) and (26), while (8) was found to degrade faster in soils at higher moisture contents, at least in part from catalytic hydrolysis. Similarly as (3), the addition of TDTMA Br to methidathion (33) on a soil thin layer did not show any marked change of degradation under natural sunlight (Sánchez et al. 2005). Because light attenuation is significant on the soil surface, the movement of pesticide molecules in soil is considered to control the extent of photodegradation (Katagi 2004). Under natural sunlight, the degradation rate of atrazine (13) in a soil thin layer was greater in wet soil than dry (Gong et al. 2001; Xiaozhen et al. 2005). A larger photic depth was estimated, especially in the presence of SDBS, because of the enhanced movement of (13) in soil by solubilization. Difference of formulation affected the photodegradation of napropamide (*N*,*N*-diethyl-2-(1-naphthalenyloxy)propanamide) (Stanger and Vargas 1984). On a glass plate the herbicide in EC formulation was photodegraded much faster under sunlight than that in a wetted powder (WP), while a slightly faster degradation in WP formulation was observed on soil.

In contrast to photodegradation on solid surfaces, more investigations have been conducted in solution. The butyl ester of 2,4-D (26) in hexane was photodegraded slightly faster than its formulation with formation of the dechlorinated derivative (Que Hee et al. 1979). Butyl 2-hydroxyphenylacetate in trace amounts was considered to be formed via photo-Fries rearrangement under the cage effect of micelles, as the formulation contained about 5 % (w/w) surfactants. When the adjuvants in formulation contain aryl compounds, their photosensitization is considered to pro-

Table 7 Effect of surfactant or formulation on photolysis of pesticides

No.	Pesticide/structure	Light source, filter Temp, reaction medium Surfactant or formulation	$[DT_{50}^{o}]$ DT_{50}^{s} [a]	Reference
Deposit or film on glass				
21	Chinomethionat 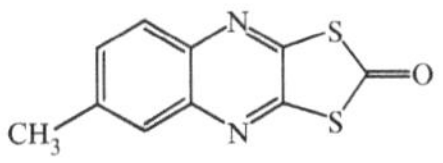	15W Fluorescence black light, none 20°C, film on glass Tween 60[b], 80, 85; 20-fold (w/w)	[>>8 hr] 4–8, ~1, ~1 hr	Nutahara and Murai (1984)
22	Chlorimuron 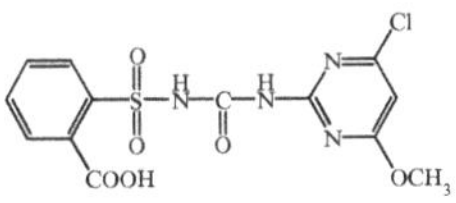	Sunlight (Aug), borosilicate glass 38±3°C, deposit on glass 0.1% Tween 80, Triton X-100	[10.9 d] 11.2, 10.7 d	Thomas and Harrison (1990)
23	Metsulfuron 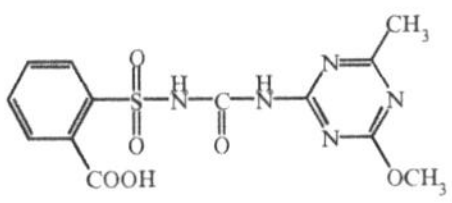	Sunlight (Aug), borosilicate glass 38±3°C, deposit on glass 0.1% Tween 80, Triton X-100	[9.5 d] 8.0, 13.2 d	Thomas and Harrison (1990)
24	Sulcotrione 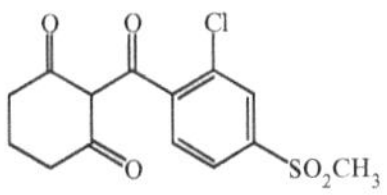	1.1 kW Xe lamp (solar simulator), none 35°C, carnauba wax film (0.8 mm) Mikado® formulation	[3.2 hr] 1.4 hr	Halle et al. (2000)
25	Azadirachtin-A 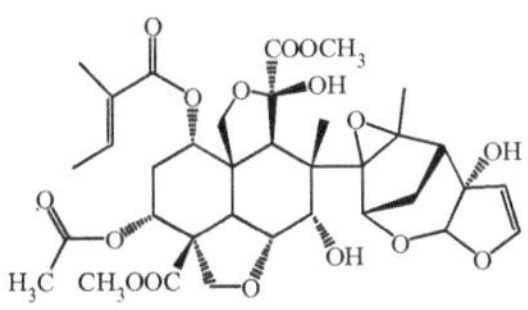	UV light at 254 nm, none NA, film on Petri dish 1:1 (w/w), 9 nonionic surfactants	[48 min] 31–94 min	Johnson and Dureja (2002)
Solution				
26	2,4-D (acid) 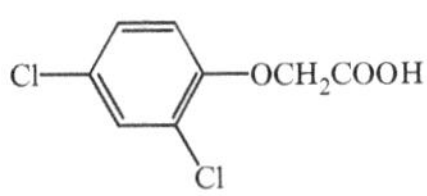	Hg lamp at 253.7 nm, none NA, distilled water at pH 3 10 mM SDS, Tween 80, HDTMA Br, Brij 35	[0.96 hr] 0.38–0.96 hr	Kwan and Chu (2005)
	2,4-D (*n*Bu ester)	UV at 300 nm, Pyrex 27°C, hexane EC formulation (5% surfactant)	[3.9 hr] 6.7 hr	Que Hee et al. (1979)

(continued)

Table 7 (continued)

No.	Pesticide/structure	Light source, filter Temp, reaction medium Surfactant or formulation	$[DT_{50}^{o}]$ DT_{50}^{s} [a]	Reference
	2,4-D (Me ester)	Sunlight (midsummer), none NA, distilled water 10 mM HDTMA Br, 50mM SDS	[62 d] 500, 21 d	Hautala (1978)
27	Haloxyfop 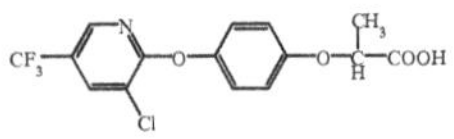	GE-RSM Sunlamp, none 19°C, distilled water 0.15% Tween 20, 1% soybean oil conc.	[~31 hr] 6, 3 hr	Harrison and Wax (1985)
1	DDT 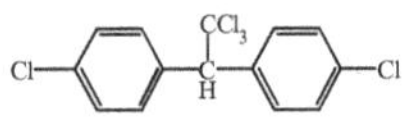	Hg lamp at 253.7 nm, none NA, distilled water 15 mM Brij-35, Brij-52, Brij-72[b]	[5.8 min] 4.9, 2.6, 4.3 min	Chu (1999)
3	Parathion 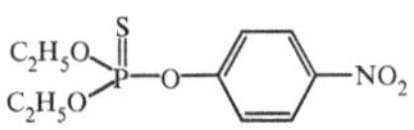	Sunlight (midsummer), none, NA, distilled water 10 mM HDTMA Br, 50mM SDS	[9.2 d] 2.9, 1.3d	Hautala (1978)
7	Coumaphos 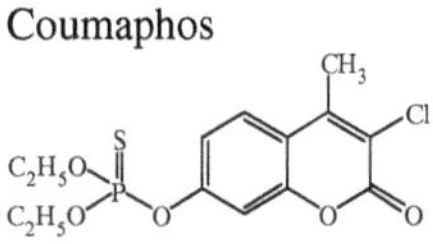	125W Hg lamp, Pyrex 30°C, phosphate buffer at pH 7.4 25 mM SDS	[1 hr] 0.89 hr	Moreno et al. (1995)
28	Bentazone 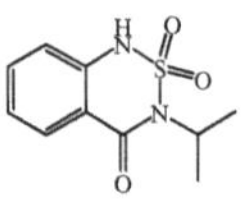	GE-RSM Sunlamp, none 19°C, distilled water 0.15% Tween 20, 1% soybean oil conc.	[~27 hr] 38, 6 hr	Harrison and Wax (1985)
8	Carbaryl 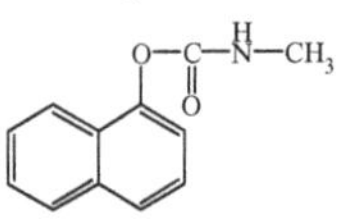	Sunlight (midsummer), none NA, distilled water 10 mM HDTMA Br, 50mM SDS	[11 d] 1.2, 9.6 d	Hautala (1978)
22	Chlorimuron 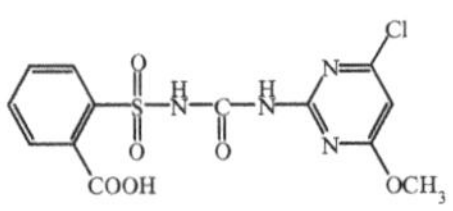	Sunlight (Aug), borosilicate glass 38±3°C, HPLC grade water at pH 6.3 0.1% Tween 80, Triton X-100	[5.8 d] 3.7, 2.7 d	Thomas and Harrison (1990)
23	Metsulfuron 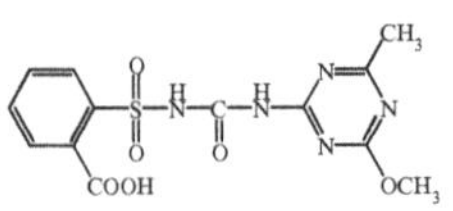	Sunlight (Aug), borosilicate glass 38±3°C, HPLC grade water at pH 6.3 0.1% Tween 80, Triton X-100	[15.7 d] 2.9, 1.5 d	Thomas and Harrison (1990)

(continued)

Table 7 (continued)

No.	Pesticide/structure	Light source, filter Temp, reaction medium Surfactant or formulation	[DT_{50}^{o}] DT_{50}^{sa}	Reference
29	Rimsulfuron 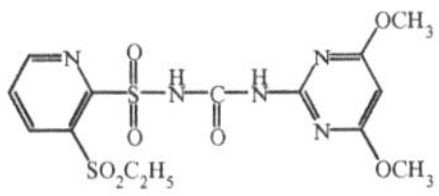	1.1kW Xe lamp (solar simulator), none 25°C, acetonitrile 25% WG formulation	[79.5 hr] 42.3 hr	Scrano et al. (1999)
30	Clethodim 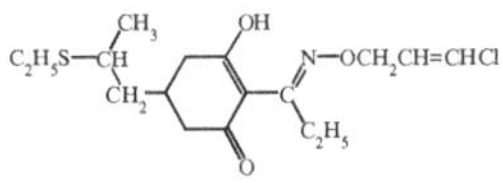	Sunlight (Oct), none 22±1°C, 0.1M K_2HPO_4 buffer at pH 7 1% 5 adjuvants	[>8 hr] 0.3–1.3 hr	Falb et al. (1990)
31	Hexaconazole 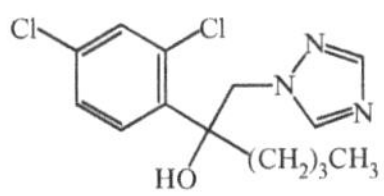	125W Hg lamp, water NA, distilled deionized water Anvil® formulation (ai, 10 ppm)	[13.9 min] 34.7 min	Santoro et al. (2000)
32	Imidacloprid	125W Hg lamp, Solidex filter (>280 nm) NA, HPLC grade water Confidor® formulation	[0.72 hr] 2.1 hr	Wamhoff and Schneider (1997)
Soil surface				
26	2,4-D (Me ester) 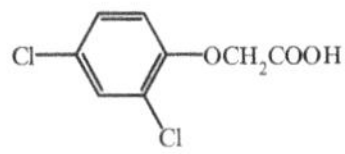	450W Hg lamp, Pyrex NA, 4 US dry soil thin-layer (30 μm) 10% (w/w) HDTMA Br, SDS	[3.6–6.7 d] 3.7–4.3, 3.0–10.1 d	Hautala (1978)
8	Carbaryl 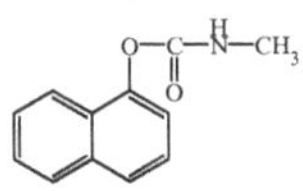	450W Hg lamp, Pyrex NA, 4 US dry soil thin-layer (30 μm) 10% (w/w) HDTMA Br, SDS	[4.0–10.5 d] 10.5–14.9, 10.5–11.4 d	Hautala (1978)
3	Parathion 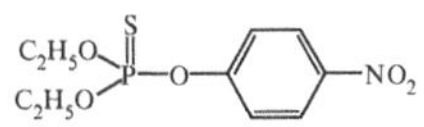	450W Hg lamp, Pyrex NA, 4 US dry soil thin-layer (30 μm) 10% (w/w) HDTMA Br, SDS	[2.2–2.9 d] 3.8–5.6, 1.6–4.5 d	Hautala (1978)

(continued)

Table 7 (continued)

No.	Pesticide/structure	Light source, filter Temp, reaction medium Surfactant or formulation	$[DT_{50}^{o}]$ DT_{50}^{s} [a]	Reference
33	Methidathion 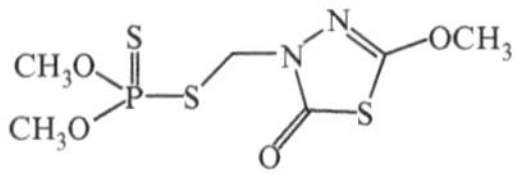	Sunlight (Jul-Aug), borosilicate glass 15–38°C, soil thin-layer (0.9 mm) TDTMA Br (10 × cmc)	[1.7 d] 1.6 d	Sánchez et al. (2005)
13	Atrazine 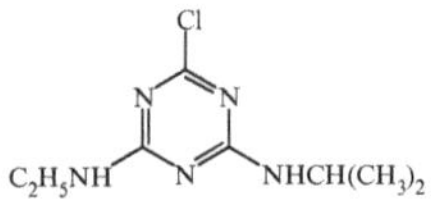	1 kW Hg lamp, water 28±2°C, soil thin-layer (0.2–0.5 mm) SDBS (20 mg/kg soil)	[5.5 min] 5.2 min	Gong et al. (2001)

Brij-52, $C_{16}H_{33}(OCH_2CH_2)_2OH$; Brij-72, $C_{18}H_{37}(OCH_2CH_2)_2OH$; r.t., room temperature; NA, not available; None, without filter.

[a] Half-life of degradation assuming pseudo-first-order kinetics with (s) or without (o) surfactant or adjuvant.

[b] Tween 60, monostearate of nonionic surfactant (d) in Fig. 2.

mote photodegradation, as demonstrated for (26) and haloxyfop (27) (Harrison and Wax 1985). Hautula (1978) has reported the increase of a quantum yield with shorter half-lives for sunlight photolysis of (26), parathion (3), and carbaryl (8), partly due to the red shift of their absorption spectra.

The photodegradation process of (26) in micelles has been studied extensively by Kwan and Chu (2005). SDS micelles did not affect the photodegradation rate, whereas the other micelles of HDTMA Br, Tween 80, and Brij 35 accelerated it at pH 3 by a factor of 2.5–5. The more favorable formation of an excimer followed by an electron transfer to give the anion radical-like species in the latter micelles resulted in the regioselective dechlorination (*o*/*p* ~ 2) at pH 3, but no selectivity was detected at pH 10, possibly because of inhibition of dechlorination at the *o*-position by the neighboring CH_2COO^- moiety. Photoinduced dechlorination was also examined for DDT (1) in three alcohol ethoxylate micelles (Chu 1999). Enhancement in the micelles was about twofold compared with aqueous photolysis via dechlorination at the CCl_3 group followed by the aryl moieties. In relation to the compartmentalization effect, the photodimerization of coumaphos (7) was investigated in SDS micelles, and more dimers were formed as the occupation number defined by the ratio of (7) molecule per micelle increased (Moreno et al. 1995).

The photochemical reactions of urea herbicides in nonionic micelles have been investigated extensively by using monuron (3-(4-chlorophenyl)-1,1-dimethylurea) as a model compound. Tanaka et al. (1979, 1981) reported an increased rate of photolysis for monuron in the micelles of nonionic alkylphenoxy ethoxylates. The solubilization in micelles eliminated ring hydroxylation but the reductive dechlorination was enhanced. Demethylation proceeded as a minor pathway, with formation of HCHO implied by the photoreaction with H_2O, and neither dechlorination nor *N*-demethylation was affected by molecular oxygen. Therefore, monuron is most likely to be solubilized in the Stern layer of micelles. Detection of the corresponding biphenyl dimers together with its reduction by addition of quenchers showed that photodegradation was initiated by homolytic cleavage of the C-Cl bond (Tanaka et al. 1982; Tanaka 1989).

Photoinduced isomerization and decarboxylation in micelles were examined for some pyrethroids. UV irradiation of cypermethrin ((*RS*)-α-cyano-3-phenoxybenzyl (1*RS*)-*cis-trans*-3-(2,2-dichlorovinyl)-2,2-dimethylcyclo-propanecarboxylate) in SDS micelles specifically gave the corresponding acid, benzaldehyde, and benzoic acid, which showed the effective hydrogen abstraction from the surfactant by the radicals formed via homolytic cleavage of the O—CH(CN) bond (Ruzo 1983). In the case of cyhalothrin, the 2-chloro-2-trifluoromethylvinyl analogue of cypermethrin, the main reactions in SDS and DDTMA Br micelles were photoinduced isomerization and decarboxylation from the greater stability of the generated radicals (Ruzo et al. 1987).

Photoreduction of nitrodiphenyl ether herbicides has been examined in micelles. By exposure to UV-A light, the several nitrodiphenyl ethers in Triton X-100 micelles were found to be reduced by the cosolubilized β-carotene (Orr and Hogan 1985). The reactive species was a nitroxide radical, as demonstrated by spin trapping technique using electron spin resonance (ESR) (Piekarski et al. 1990). The broadening of the asymmetrical ESR signal in the higher field together with a redshift of absorption spectrum indicated that the nitrodiphenyl ether is dissolved near the core of SDS and HDTMA Br micelles. Photodegradation of sulfonylurea herbicides having a high water solubility is also accelerated by nonionic surfactants. Both Tween 80 and Triton X-100 markedly enhanced the sunlight photodegradation of chlorimuron (22) and metsulfuron (23) with a larger extent by the latter surfactant, possibly because of sensitization of an aryl moiety of the surfactant (Thomas and Harrison 1990). Similar photosensitization has been reported for rimsulfuron (29) in formulation (Scrano et al. 1999). The effect of various adjuvants on the photostability of cyclohexene oxime herbicides has been well examined from the aspect of preparing a stable formulation (Falb et al. 1990; Hazen and Krebs 1992; McInnes et al. 1992). Factors determining photostability of pesticides in formulation cannot be easily identified because of confidentiality of the adjuvants, and either enhancement or retardation of photodegradation in formulation was reported (Mekkaoui et al. 2000; Santoro et al. 2000; Wamhoff and Schneider 1997).

VI Effects on Mobility and Biodegradation in Soil

A *Adsorption and Desorption*

There are many possible processes involved in the adsorption and desorption of a chemical to soil in the presence of surfactant (see Fig. 5). In the pore water of soil, surfactant can be present as a monomer or micelle depending on its concentration.

Furthermore, the surfactant would be adsorbed onto the soil surface as a monomer, hemimicelle, or admicelle. Adsorption of a chemical to soil would be reduced when the chemical is solubilized to micelles in a water phase, whereas it increases by solubilization to hemimicelles and admicelles on soil (Krogh et al. 2003; Pennell et al. 2003). Therefore, the complex interactions of a chemical with each species in both the water and soil phase together with the concentration of surfactant should be taken into account.

Simple Organic Chemicals

Many investigations have been conducted from the aspect of remediation of soil contaminated with PAHs. Aronstein et al. (1991) reported the enhanced desorption of biphenyl and phenanthrene from mineral soil by using two alcohol ethoxylates having the HLB values around 11, but the effect was minimal for organic muck soil. The desorption of phenanthrene and pyrene from loamy sand soil was found to increase markedly by a factor of more than 10 when nonionic surfactants were used, but SDS showed an insignificant effect, possibly because of its lower affinity to soil surface (Cheng and Wong 2006a). The desorption K_d values of two PAHs mostly decreased on the order of Tween 80 > Triton X-100 > Brij 35 > SDS in the presence of 150 – 300 mg surfactant/L. The addition of dissolved organic matter (DOM) enhanced desorption of these PAHs by Tween 80 several times (Cheng and Wong 2006b). No additive effect of DOM in desorption may imply some interactions between DOM and Tween 80.

In parallel with the experimental approach, the effect of surfactant on adsorption and desorption of a chemical has been theoretically investigated. Edwards et al. (1992) indicated that a higher concentration of surfactant in a soil–water system is necessary to show the same surface tension in water. When the Freundlich isotherm is assumed for adsorption of surfactant on soil, the apparent fraction of soil organic carbon content can be expressed as $f_{oc}^* = f_{oc} + (K_F \cdot C_{mon}^{1/n}) \cdot f_c$, where f_{oc}, K_F, C_{mon}, and f_c are the organic carbon content of soil, Freundlich adsorption coefficient of surfactant, monomer concentration, and weight fraction of C in surfactant, respectively. As the solubility enhancement of a chemical at concentrations less than cmc (sub-cmc) is proportional to the surfactant concentration (see Fig. 4a), the apparent linear adsorption coefficient K_d^* of a chemical can be written as $K_d^* = K_d\,[1 + (K_F \cdot C_{mon}^{1/n} \cdot f_c / f_{oc}) / [1 + K_{mon} \cdot C_{mon}]$, where K_d and K_{mon} are the adsorption coefficient in the absence of surfactant and the association constant between a chemical and monomer, respectively. In the case of nonionic Triton X-100, the K_d^* value of pyrene gradually decreases with an increase of surfactant concentration at sub-cmc levels. At higher concentrations of surfactant above cmc, the similar process including solubilization of a chemical to micelles greatly reduces the adsorption coefficient (Di Cesare and Smith 1994; Edwards et al. 1994).

The desorption process of PAHs from coal tar-contaminated soil has been examined by Yeom et al. (1996). The percentage of solubilization by nonionic surfactants was well explained by partition of PAHs to micelles, but the dissolution rate of phenanthrene from the contaminated soil was much larger than that from the crystalline phenanthrene. Therefore, sorption of surfactant, its penetration into soil, and swelling of soil matrix followed by matrix diffusion of a chemical were considered to be the important processes in enhanced desorption. These effects are considered to be highly dependent on the chemical class of surfactant. Jones-Hughes and Turner (2005) reported an inhibitory effect of SDS on the sorption of phenanthrene to estuarine sediment but a significant enhancement of sorption in the presence of HDTMA Br. It was considered that nonspecific adsorption of SDS to negatively charged sediment competed with phenanthrene for access to hydrophobic sites of the sediment surface but that the electrostatic adsorption of HDTMA Br gave additional sorption sites for phenanthrene. This sorption process of PAHs to the adsorbed surfactants has been examined for SDS and Tween 80 by using kaolinite clay (Ko et al. 1998). The distribution coefficients (K_D) of phenanthrene and naphthalene between clay and water gradually increased with the SDS concentration up to its cmc, but decreased afterward for reasons of more favorable partition to micelles in a water phase. Similar profiles were observed for Tween 80, but the maximum K_D values were obtained much above its cmc. These differences are considered to stem from the different adsorption characteristics of two surfactants. Because the sorption of a chemical to soil or sediment proceeds via not only partition to their organic matter fractions but also by adsorption by van der Waals or electrostatic forces, the effect of surfactant on adsorption to clay and minerals has also been investigated. Theng et al. (1998) examined the partition of phenanthrene to the complex of TDTMA Br and montomotillonite (nearly 1:1, w/w) where the surfactant molecules were intercalated into the clay as they formed bilayers with the $(CH_2)_2$ chains lying parallel, as evidenced by X-ray diffraction patterns and ^{13}C-NMR. Magic-angle spinning NMR showed that phenanthrene and TDTMA Br were intimately associated in the clay interlayer.

The adsorption of small organic molecules such as benzene and polychloroethylene derivatives to soils has been studied in the presence of alkyltrimethylammonium (Lee et al. 1989). Significant enhancement of adsorption following the linear isotherm was observed. The longer the alkyl chain of the surfactant, the higher adsorption to soils, indicating partition of the organic chemicals to the surfactants sorbed on soil. When the linear isotherm is assumed for adsorption of a chemical to soil with the coefficient of K_d, the apparent adsorption coefficient (K_d^*) at the surfactant concentration above cmc is expressed below (Lee et al. 2000):

$$K_d^* = K_d[1+f_{sf}(K_{sf} / K_d)] / (1 + X_{mon} \cdot K_{mon} + X_{mic} \cdot K_m),\ X_{mic} = X - cmc$$

X is the total concentration of surfactant in a mass/mass unit, and the suffices "mon" and "mic" mean monomer and micelle, respectively. K_{sf} and f_{sf} are the partition coefficient of surfactant on soil and mass fraction of the sorbed surfactant in soil, respectively. The K_d^*/K_d values in the presence of nonionic surfactants including Triton X-100 were larger than 1 for benzene derivatives and gradually increased to plateau (Lee et al. 2000, 2006). When SDBS and domiphen bromide $[CH_3(CH_2)_{10}CH_2(CH_3)_2N^+(CH_2)_2OPh\ Br^-]$ were used, the K_d^*/K_d values showed maxima at the surfactant concentration of 2–5 × cmc because of their solubilization to micelles in a water phase (Lee et al. 2004). The modification of soils with HDTMA Br also enhanced the adsorption of benzene derivatives (Boyd et al. 1988; Gao et al. 2001; Jaynes and Boyd 1991). In the case of positively charged aluminum oxide, the sorbed SDS and nonionic surfactants have been reported to enhance the adsorption of many types of organic chemicals (Nayyar et al. 1994; Park and Jaffé 1993; Valsaraj 1992).

Pesticides

Soil adsorption parameters of some pesticides based on the Freundlich isotherm together with the effect of adjuvants including surfactants are summarized in Table 8. The reported effects are highly dependent not only on the chemical class of adjuvants and their concentration but also those of pesticides. Generally, the addition of an anionic surfactant reduces the K_F value by means of its competitive adsorption on soil and solubilization to micelles, but a slight increase of adsorption is sometimes observed (Ye 2003). Either nonionic or cationic surfactant tends to increase the K_F value by the partition of pesticide to the hydrophobic parts of the soil-sorbed surfactants, especially with a larger extent for the latter surfactant, but it sometimes decreases at a higher concentration than cmc from solubilization to micelles in an aqueous phase (Huggenberger et al. 1973; Iglesias-Jiménez et al. 1996, 1997; Singh et al. 2000, Sánchez et al. 2003b; Sun and Inskeep 1995). The partition of pesticide to the hydrophobic sites consisting of the alkyl chains of nonionic surfactants on soil has been confirmed by the positive correlation of the maximum adsorption amounts of moderately hydrophobic fungicides with the total hydrophobic carbon number of surfactant (Steurbaut 1994). Locke et al. (2002) reported that the hydrophobic norflurazon (46) is adsorbed more to soil in the presence of nonionic Triton X-77, but the K_d values of more hydrophilic triazine herbicides are negatively correlated with the carbon number of the surfactants. The ionizable 2,4-D (26) was adsorbed less to soil when a surfactant having a lower HLB value was present, showing the importance of electrostatic interactions (Amonette and O'Conner 1990).

Because the usual pesticide formulation contains an adjuvant consisting of surfactant and other components, the aforementioned effects are masked in many cases. The adsorption profiles in soil, sediment, and clay are almost the same between an a.i. and its formulation (Beestman and Deming 1976; Beigel and Barriuso 2000; Ekler 1988; García-Ortega et al. 2006; Singh et al. 2002a), and the effect of a formulation type is usually insignificant (Chaplain et al. 2001; Cotterill 1988; Mingelgrin and Yaron 1973). However, in the case of urea and carbamate herbicides, less adsorption to river sediment was reported for the formulations (Margoum et al. 2006). The effect of surfactant on pesticide adsorption is also known to be dependent on soil characteristics. The addition of Triton X-77 enhanced the adsorption of primisulfuron (43) to soils with low organic carbon (oc)

Table 8 Effect of surfactant and formulation on soil adsorption and desorption of pesticide

No.	Pesticide/structure[a]	Soil (oc, CEC, pH)[b] Surfactant[c]	K_F[d]	n[d]	Reference
26	2,4-D (acid) (a)	Silt loam (1.74, 19.0, 5.9)*	1.90	1.11	Amonette and O'Connor (1990)
	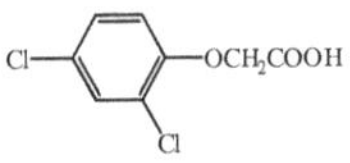	+ Triton N-150[e] (0.5%, w/v)	1.74	1.14	
		+ Triton N-60[e] (0.5%, w/v)	1.69	1.10	
		+ Triton X-35[e] (0.5%, w/v)	1.22	1.14	
		Clay loam (0.46, 31.1, 7.7)*	0.21	1.16	
		+ Triton N-150 (0.5%, w/v)	0.10	1.11	
		+ Triton N-60 (0.5%, w/v)	0.18	1.04	
		+ Triton X-35 (0.5%, w/v)	0.05	0.93	
34	Lindane (d)	Sandy loam (7.85, 4.98, 4.0)	52.4	L	Quintero et al. (2005)
	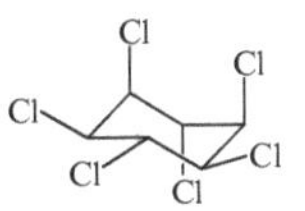	+ Tween 80 (5 × cmc)	29.3	L	
		+ Triton X-100 (5 × cmc)	23.9	L	
		+ SDS (5 × cmc)	48.7	L	
35	Heptachlor (a)	Sandy (1.39, 3.4, NA)	121	L	Lee et al. (2000)
	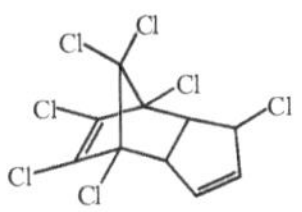	+ Triton X-100 (68.4 ppm)	145	L	
		(312 ppm)	82.4	L	
1	DDT (a)	Silt loam (0.10, NA, NA)	456	L	Sun and Inskeep (1995)
	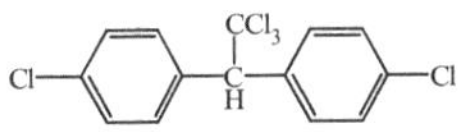	+ Triton X-100 (64 ppm)	856	L	
		(613 ppm)	43.3	L	
36	Dichlorvos (a)	Spanish soil (0.27, 23.5, NA)	NA	NA	Sánchez-Camazano and Sánchez-Martin (1994)
	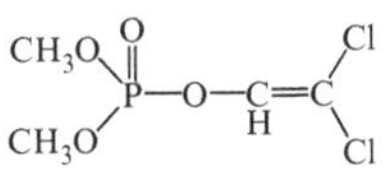	+ HDTMA (74% CEC equiv)	75.9	0.79	
		Spanish soil (2.90, 10.9, NA)	NA	NA	
		+ HDTMA (4% CEC equiv)	47.9	0.60	
37	Acephate (a)	Spanish soil (0.45, 9.4, 7.5)	0.72	0.93	Iglesias-Jiménez et al. (1996)
	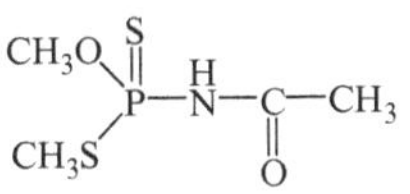	+ TDTMA Br (0.1 × cmc)	0.39	1.12	
		(cmc)	0.14	1.26	
		(20 × cmc)	0.93	1.06	
		+ SDS (cmc)	0	NA	
		+ Tween 80 (cmc)	0.85	0.97	
		(20 × cmc)	0.64	1.02	
33	Methidathion (a)	Spanish soil (0.92, 7.9, 8.5)	2.70	1.25	Sánchez et al. (2003b)
	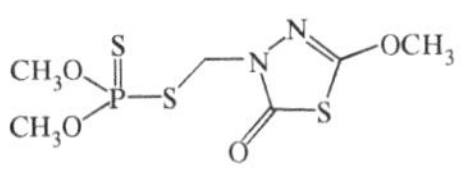	+ Tween 80 (10 × cmc)	1.52	0.97	
		+ LAS[e] (10 × cmc)	1.98	1.15	
		+ TDTMA Br (10 × cmc)	44.8	0.91	

(continued)

Table 8 (continued)

No.	Pesticide/structure[a]	Soil (oc, CEC, pH)[b] Surfactant[c]	K_F[d]	n[d]	Reference
38	Diazinon (a)	Spanish soil (0.45, 9.4, 7.5)	9.95	0.91	Iglesias-Jiménez
	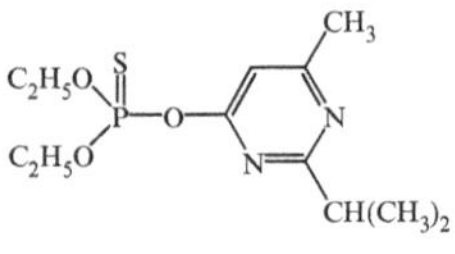	+ TDTMA Br (0.1 × cmc)	8.09	1.01	et al. (1996)
		(cmc)	12.96	0.91	
		+ TDTMA Br (20 × cmc)	15.39	0.88	
		+ SDS (cmc)	5.44	0.90	
		(20 × cmc)	0	NA	
		+ Tween 80 (cmc)	9.70	0.95	
		(20 × cmc)	14.2	0.93	
39	Azinphos-methyl (a)	Spanish soil (0.27, 23.5, NA)	NA	NA	Sánchez-
	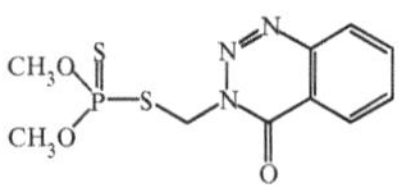	+ HDTMA (74% CEC equiv)	933	0.84	Camazano
		Spanish soil (2.90, 10.9, NA)	NA	NA	and Sánchez-
		+ HDTMA (4% CEC equiv)	75.9	1.06	Martin (1994)
9	Carbofuran (a)	Sandy loam (0.33, 9.0, 8.8)	22.9	1.08	Singh et al.
	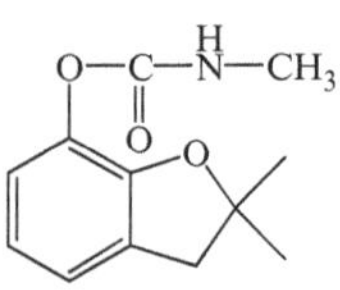	+ HDTMA Br (0.5 × cmc)	41.68	1.25	(2000)
		(2.0 × cmc)	7.76	0.96	
		+ Tween 20 (0.5 × cmc)	28.8	1.06	
		(2.0 × cmc)	3.16	0.79	
		+ SDS (0.5 × cmc)	5.49	0.79	
		(2.0 × cmc)	1.20	0.68	
40	Chlorotoluron (a)	UK soil (4.1, NA, 5.1)*	NA	NA	Cotterill (1988)
	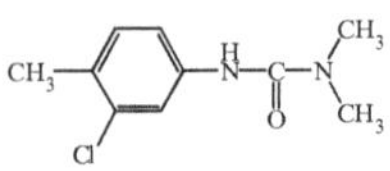	50% suspendable concentrate	8.67	1.19	
		80% wettable powder	7.41	1.15	
41	Diuron (a)	Clay loam (0.79, NA, NA)	NA	NA	Chaplain et al.
	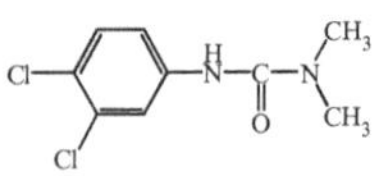	+ flowable formulation, 5 μm	2.05	L	(2001)
		or 1 μm grain size of soil	1.98	L	
42	Linuron (a)	Spanish soil (0.45, 9.4, 7.5)	2.29	NA	Iglesias-Jiménez
	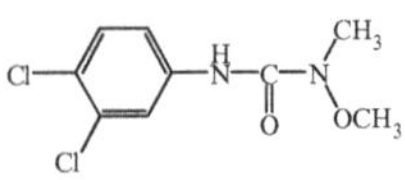	+ TDTMA Br (9.1mgC/g	72.7	NA	et al. (1997)
		soil)	0.42	NA	
		+ SDS (9.1mgC/g soil)			
	Linuron (d)	Sandy loam (4.22, NA, 5.2)	41.8	0.08	Sánchez-
		+ LAS (1.5 × cmc)	38.2	0.14	Camazano
		+ Tween 80 (10 × cmc)	40.2	0.13	et al. (2003)
		+ Triton X-100 (10 × cmc)	25.1	0.57	
		+ Brij 35 (10 × cmc)	37.7	0.21	
		+ Tergitol NP-10 (10 × cmc)	37.8	0.21	

(continued)

Table 8 (continued)

No.	Pesticide/structure[a]	Soil (oc, CEC, pH)[b] Surfactant[c]	K_F[d]	n[d]	Reference
43	Primisulfuron (a)	Sandy clay loam (0.66, NA, 4.5)*	0.55	L	Werkheiser and Anderson (1996)
	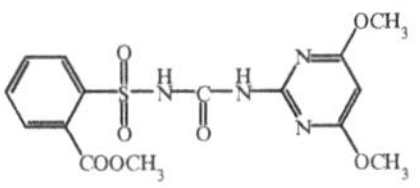	+ Triton X-77[e] (0.25%, w/v)	0.88	L	
		Clay (1.67, NA, 6.5)*	0.18	L	
		+ Triton X-77 (0.25%, w/v)	0.14	L	
44	Acetochlor (a)	Chinese soil (1.5, 84.0, 8.4)	4.339	1.57	Ye (2003)
	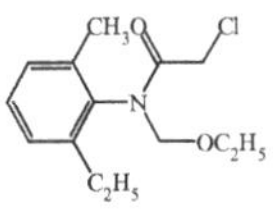	+ SDBS (40 mg/L)	4.303	1.61	
		(2.24 g/L)	5.489	1.27	
45	Metalaxyl (a)	Sandy loam (0.99, NA, 5.1)	14.1	0.91	Steurbaut (1994)
	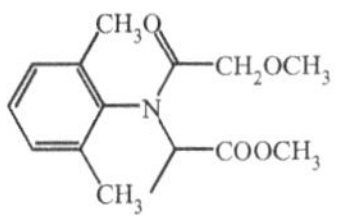	+ NP 6EO[e] (NA)	10.3	0.86	
		+ NP 18EO[e] (NA)	12.1	0.90	
46	Norflurazon (a)	Silt loam (0.68, 12.4, 6.3)	3.88	L	Locke et al. (2002)
	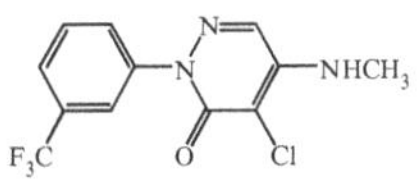	+ Triton X-77 (0.25%)	4.66	L	
		Clay (1.69, 43.7, 6.0)	8.76	L	
		+ Triton X-77 (0.25%)	9.82	L	
16	Iprodione (a)	Sandy loam (0.99, NA, 5.1)	26.1	1.04	Steurbaut (1994)
	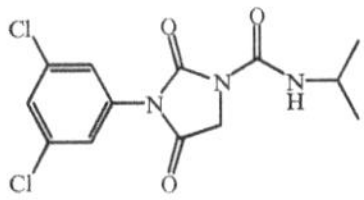	+ NP 6EO (NA)	13.7	0.92	
		+ NP 18EO (NA)	17.4	1.12	
17	Vinclozolin (a)	Sandy loam (0.99, NA, 5.1)	36.2	0.86	Steurbaut (1994)
	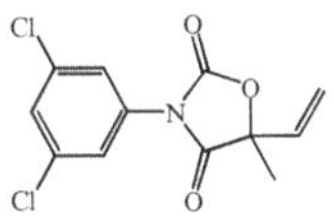	+ NP 6EO (NA)	13.9	1.06	
		+ NP 18EO (NA)	14.3	1.10	
47	Captan (a)	Sandy loam (0.99, NA, 5.1)	63.7	1.25	Steurbaut (1994)
	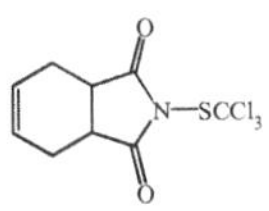	+ NP 6EO (NA)	27.2	1.10	
		+ NP 18EO (NA)	38.5	1.19	
48	Triticonazole (a)	Loamy clay (1.04, NA, 8.2)	4.61	0.87	Beigel and Barriuso (2000)
	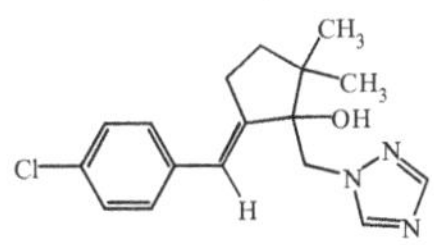	+ Real® blank formulation	4.09	0.92	
		+ Real® & Soprophor FLK[e]	4.02	0.89	
		+ Real® & MNS90[e]	4.55	0.92	

(continued)

Table 8 (continued)

No.	Pesticide/structure[a]	Soil (oc, CEC, pH)[b] Surfactant[c]	K_F[d]	n[d]	Reference
49	Propiconazole (a) 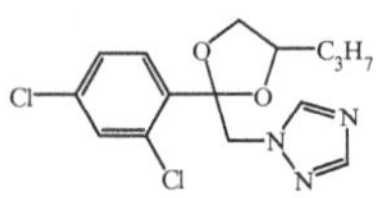	Sandy loam (0.99, NA, 5.1)	25.9	1.27	Steurbaut (1994)
		+ NP 6EO (NA)	12.8	2.08	
		+ NP 18EO (NA)	15.0	1.10	
13	Atrazine (a) 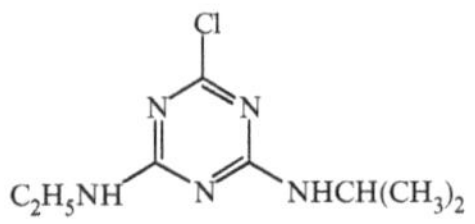	Spanish soil (0.45, 9.4, 7.5)	0.68	0.81	Iglesias-Jiménez et al. (1996)
		+ TDTMA Br (0.1 × cmc)	0.65	0.86	
		(cmc)	0.75	0.78	
		(20 × cmc)	2.46	0.90	
		+ SDS (cmc)	0.41	0.93	
		(20 × cmc)	0	NA	
		+ Tween 80 (cmc)	0.73	0.92	
		(20 × cmc)	0.86	0.85	
		Silt loam (0.68, 12.4, 6.3)	1.75	L	Locke et al. (2002)
		+ Triton X-77 (0.25%)			
13	Atrazine (a)	Clay (1.69, 43.7, 6.0)	4.59	L	Locke et al. (2002)
		+ Triton X-77 (0.25%)	4.28	L	
		River sediment (2.06, 16.3, 7.76)	1.71	L	
		+ SDBS (10 ppm)	0.45	L	Tao et al. (2006)
		+ HDTMA Br (10 ppm)	1.10	L	
		+ Span 20[e)] (10 ppm)	2.77	L	
	Atrazine (d)	Spanish soil (0.81, NA, 7.6)	4.58	0.39	Sánchez-Camazano et al. (2000b)
		+ SDS (0.75 × cmc)	5.04	0.33	
		(1.5 × cmc)	6.50	0.25	
		Spanish soil (5.97, NA, 4.7)	33.2	0.09	
		+ SDS (0.75 × cmc)	29.4	0.12	
		(1.5 × cmc)	20.4	0.34	
50	Cyanazine (a) 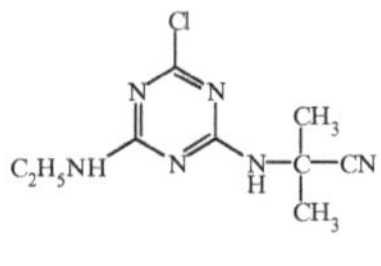	Silt loam (0.68, 12.4, 6.3)	1.75	L	Locke et al. (2002)
		+ Triton X-77 (0.25%)	1.27	L	
		Clay (1.69, 43.7, 6.0)	5.59	L	
		+ Triton X-77 (0.25%)	4.50	L	
51	Ethofumasate (a) 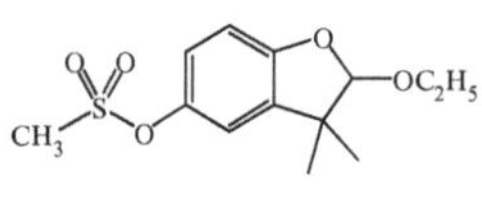	Spanish soil (0.45, 9.4, 7.5)	0.89	1.03	Iglesias-Jimenez et al. (1996)
		+ TDTMA Br (0.1 × cmc)	1.37	0.91	
		(cmc)	2.59	1.31	
		(20 × cmc)	136.8	1.09	
		+ SDS (cmc)	1.59	0.90	
		(20 × cmc)	0	NA	
		+ Tween 80 (cmc)	1.52	0.93	
		(20 × cmc)	1.73	0.96	

(continued)

Table 8 (continued)

[a]"a" and "d" in the parentheses mean adsorption and desorption, respectively.
[b]oc, organic carbon (%), corrected by a factor of 1/1.724 from organic matter content; CEC, cation-exchange capacity (meq/100 g soil); NA, not available; *, the usage of $CaCl_2$ solution as an aqueous phase.
[c]cmc, critical micelle concentration; nEO, polyethoxylated chain with a unit number of n.
[d]Freundlich sorption coefficient (L/kg) and the constant estimated by batch equilibration technique; $x/m = K_F \cdot C^{1/n}$, where x and m are the amounts of pesticide and soil and C is the aqueous concentration of pesticide. If the n column is 'L', K_F means the sorption coefficient in linear isotherm.
[e]NP 6EO & Triton N-60, $C_9H_{19}C_6H_4(OCH_2CH_2)_6OH$; NP 18EO, $C_9H_{19}C_6H_4(OCH_2CH_2)_{18}OH$; LAS, linear alkylbenzenesulfonate; Triton X-35, $C_8H_{17}C_6H_4(OCH_2CH_2)_3OH$; Triton X-77, mixture of alkylarylpolyethylene glycols, free fatty acids, and isopropanol; Triton N-150, $C_9H_{19}C_6H_4(OCH_2CH_2)_{15}OH$; Span 20, sorbitan monolaurate; Soprophor FLK, phosphate tristyrylphenol ethoxylate; MNS90, sodium alkylnaphthalenesulfate.

contents (<1%) but caused slightly less adsorption for soils with higher oc (Werkheiser and Anderson 1996). In addition to soil characteristics, the ability of surfactant to disperse soil matrix and then increase the sites for adsorption is also considered to be an important factor (Tao et al. 2006).

The desorption of pesticide from soils is also affected by surfactant. Both Triton X-100 and Tween 80 at higher concentrations than cmc could enhance the desorption of hydrophobic lindane (34) from soil but an insignificant effect was observed by SDS, which seemed to depend on the hydrophobicity of surfactant (Quintero et al. 2005). In contrast, the efficiency of desorption of more hydrophilic aldicarb (60) from soil by Triton X-100 above cmc was less than those of HDTMA Br and SDS (Xu et al. 2006). Mata-Sandoval et al. (2002) reported insignificant desorption of pesticides from soil by Triton X-100 below cmc, but the enhanced desorption above cmc was observed in the order of trifluralin (15) > coumaphos (7) >> atrazine (13). Similar profiles were obtained for rhamnolipid biosurfactant. This order is in agreement with that of log P and, therefore, the above different effects in desorption are likely to originate from the affinity between pesticide and surfactant. When the Freundlich isotherm is compared between adsorption of pesticide to and its desorption from soil, the n values are different because of hysteresis. To quantify the hysteresis, the H value defined by n (adsorption) / n_d (desorption) is utilized and usually increases with soil organic matter content. The greater the desorption efficiency of surfactant at a concentration higher than cmc, the lesser the H value. These profiles have been confirmed by the adsorption and desorption studies of atrazine (13) and linuron (42) for several soils with the desorption efficiency proportional to the surfactant concentration (Sánchez-Camazano et al. 2000a,b). When soil applied with pesticide is aged, the desorption profiles are known to change compared with those from a fresh soil, partly by the entrapment of pesticide in soil matrix (Katagi 2006). Triton X-100 has been shown to desorb trifluralin (15) from freshly treated and aged soils very similarly, with the extent proportional to its concentration (Mata-Sandoval et al. 2000). Therefore, Triton X-100 is considered to access the interior of soil particles and release tightly adsorbed pesticide. A similar ability for releasing pesticides from the aged soils was reported for several nonionic surfactants, and their extent gradually decreased with a period of aging (Ahmad et al. 2004; Rodriguez-Cruz et al. 2006).

Clay mineral is one of the main components in soil, and surfactant can be either adsorbed onto its surface or intercalated into its interlayers. The adsorption of atrazine (13) to sodium-saturated smectite was influenced by nonionic Brij-35 surfactant (Chappell et al. 2005). The K_F values decreased with Brij 35 concentration up to 2.1 g/L but increased afterward. Based on the X-ray diffraction data, Brij-35 was considered to first compete with (13) for intercalation to the clay but at the higher concentration the admicelles on the clay surface would solubilize (13). The cationic surfactant DDTMA Br has been shown to replace paraquat (1,1-dimethyl-4,4′-bipyridinium) intercalated into laponite clay, as evidenced by batch equilibration method and X-ray diffraction measurements (Esumi et al. 1998). The character of clay sorbing surfactant molecules into its interlayer has been applied extensively to prepare organoclay formulations of hydrophilic pesticides to improve their mobility in soil. 2,4-D (26) is an ionizable herbicide that is scarcely adsorbed to montmorillonite clay, but treatment of the clay with HDTMA Br at a concentration equivalent to cation-exchange capacity (CEC) greatly enhances its adsorption (Hermosín et al. 2006). The desorption process showed insignificant hysteresis, indicating weak hydrophobic interactions between (26) and HDTMA Br intercalated into the clay.

Enhanced adsorption of organophosphorus pesticides has been examined using montmorillonite modified by quaternary ammonium surfactants. Rodríguez-Gonzalo et al. (1993) reported increase of K_F values for parathion (3) and its oxon (4) by 1.0 and 0.3, respectively, in a logarithm unit when the natural montmorillonite was treated with 84% CEC equivalent TDTMA Br. The larger affinity of (3) to the organoclay than (4) corresponds to the more hydrophobicity of (3), which is in accordance with the adsorption mechanism of partition to the intercalated surfactants. The adsorption isotherm of azinphos-methyl (39) was found to depend on the extent of a countercation exchange in montmorillonite by HDTMA Br and DDTMA Br (Sánchez-Camazano and Sánchez-Martin 1994). In the case of HDTMA Br, the 50% exchanged clay showed a linear isotherm, but 99% and 25% exchange resulted in type-L and type-S isotherms, respectively. The log K_F values of (39) increased with the alkyl chain length of the surfactants, while insignificant change as compared with the untreated clay was observed for the more water-soluble dichlorvos (36). Although the water solubility is similarly high, hexazinone (59) was found to be more adsorbed to montmorillonite treated with HDTMA Br, presumably because of incorporation into the unoccupied interlayer space (Celis et al. 2002).

The structure of surfactant sorbed to clay has been reported to change the adsorption mechanism of pesticide to organoclay (Singh et al. 2004). Montmorillonite treated with phenyltrimethylammonium chloride (PTMA Cl) exhibited the S-type isotherm in the adsorption of fenamiphos (ethyl 4-methylthio-*m*-tolyl isopropylphosphoramidate), while linear adsorption was obtained for that with HDTMA Br. The adsorption of the pesticide onto the clay surface modified with PTMA Cl via side-by-side association is the controlling mechanism for the former organoclay, but partition to the hydrophobic phase of HDTMA Br is the controlling mechanism for the latter.

Interactions between pesticide and surfactant molecules have been investigated in more detail for sulfonylurea herbicides. By separating the micelles and monomers of octadecyltrimethylammonium bromide (ODTMA Br) with a dialysis bag, Mishael et al. (2002, 2003) investigated the interactions of each species with montmorillonite by X-ray diffraction and freeze-fracture electron microscopy. At the lower concentrations, the monomers are adsorbed as bilayers lying parallel to the clay interlayer, and its structure changed to pseudo-trimolecular layers as the surfactant concentration increased, together with adsorption of the micelles onto the clay surface. The interactions of sulfometuron (11) and sulfosulfuron (12) with each surfactant species resulted in enhanced adsorption to the organoclay. The UV absorption spectrum of sulfentrazone (56) in aqueous suspension of the same organoclay has shown that (56) is present in a dissociated form and, therefore, the tight association of (56) with ODTMA Br is likely to originate from both electrostatic and hydrophobic interactions (Polubesova et al. 2003). The interactions of alachlor (53) and the intercalated surfactant molecules have been studied using the infrared difference spectrum (El-Nahhal et al. 1998). Enhancement of adsorption to montmorillonite was minimal for the treatment of HDTMA Br, but the incorporation of benzyltrialkylammonium ion markedly increased the adsorption of (53) because of the π–π interactions of the phenyl rings and hydrogen bonding via the anilidic moiety. Similar interactions were reported for metolachlor (55) (Singh 2006).

B Leaching

Adsorption and desoprtion are the determining processes for the distribution of pesticide in soil and sediment systems, and the corresponding parameters such as a K_F value are estimated in an equilibrium state. However, the distribution of pesticide in the real environment is highly dependent on either diurnal water movement at a soil surface affected by precipitation or its continuous movement in a river water–sediment system (Katagi 2004, 2006). To investigate the mobility of pesticide in soil, a soil thin layer coated on a glass plate has been conveniently utilized, but a soil column leaching study is more favorably applied to examine the possibility of groundwater contamination by pesticide.

Helling (1971) reported increase of R_F values for several pesticides in thin-layer chromatography (TLC) using silty clay loam soil when developed by a 0.01% – 1% aqueous solution of Tween 20 and 80 surfactants. Similar enhanced movement has been reported by Foy (1992), but even 10% Tween 80 in a mobile phase was found not to change the R_F values of pesticides including atrazine (13) and trifluralin (15) in the case of organic soil. The solubilization of pesticide in micelles is considered to restrict the interactions between pesticide and soil surface, which might be in accordance with the larger enhancement of mobility by using nonionic surfactants having the longer ethoxylate unit (Steurbaut 1994). The presence of HDTMA Br in a mobile phase reduced the movement of organophosphorus pesticides in soil TLC by solubilization of the pesticides into admicelles and hemimicelles sorbed on soil via an ion-exchange mechanism (Sharma et al. 1985). When soil treated with HDTMA Br was used for preparation of soil TLC, a higher effect was observed (Sánchez-Camazano et al. 1995). In contrast, the usage of SDS either in preparation of soil TLC or as a mobile phase greatly increased the R_F values of pesticides, which was accounted for by solubilization into the anionic micelles less adsorbed onto the soil surface because of electrostatic repulsion. Singh and Kumar (2000) investigated concentration dependency of the surfactant effect in soil- and water-treated systems. Generally, at concentrations lower than cmc both nonionic and cationic surfactants reduced the mobility of pesticide by adsorption to sorbed surfactants or higher soil dispersion, but at above cmc they increased it by solubilization to micelles in a mobile phase.

The effects of adjuvants including surfactants on pesticide mobility in laboratory and field column leaching studies are summarized in Table 9. A glass, acrylic, or stainless steel cylinder with an internal diameter of 5–10 cm and a length of 10–50 cm is packed with soil and the top of the soil column is treated with a pesticide or its formulation. The column is eluted by water or 10 mM $CaCl_2$ solution and the eluate is periodically analyzed. After elution, the soil column separated into several sections is extracted and analyzed to examine the distribution of pesticide and its metabolites. The elution pattern, or breakthrough curve (BTC), is illustrated in Fig. 9. A tracer species such as chloride ion not retained in a soil column is used to estimate a pore volume (PV) (Fig. 9a). When pesticide moves through the column via adsorption/desorption process, the elution peak is delayed (Fig. 9b). Adjuvants possibly affect both the peak position and the eluted amount (Fig. 9c,d). Bayer (1967) examined the effects of 23 surfactants at 1% and 10% concentrations on the mobility of four urea herbicides in the soil column eluted by a simulated rainfall. As reported by soil TLC studies, both nonionic and anionic surfactants mostly increased the downward mobility of pesticide and some cationic ones markedly lessened the mobility. The concentration effect of surfactants in a mobile phase on the leaching of pesticide was found to be similar to those in soil TLC. The distribution of lindane (34) and diuron (41) in the soil column exhibited maxima at shallower depths when nonionic alcohol ethoxylates were used at lower concentrations, whereas the higher concentrations resulted in deeper distribution of the pesticides without clear maxima (Huggenberger et al. 1973).

Table 9 Effect of surfactants on mobility of pesticides in soil column leaching

No.	Pesticide/structure	Experimental conditions[a] Rate[b], surfactant	Effects on mobility	Reference
52	Aldrin 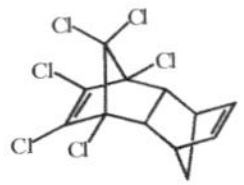	Silt loam, 8.5 cm i.d. × 15 cm, a drop / 5 sec × 17 d 10 ppm ai; linear alkylbenzene sulfonate (0.1 w/v%)	Higher laching & formation of dieldrin by surfactant but not eluted	Lichtenstein et al. (1967a)
3	Parathion 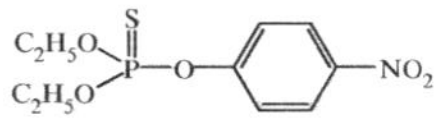	Silt loam, 8.5 cm i.d. × 15 cm, a drop / 5 sec × 17 d 10 ppm ai; linear alkylbenzene sulfonate (0.1 w/v%)	Increased persistence & higher leaching by surfactant but not eluted	Lichtenstein et al. (1967a)
53	Methidathion 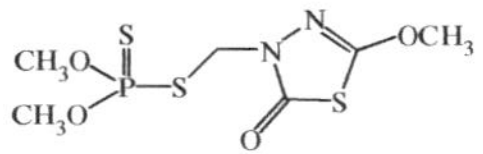	Silt loam, 5.4 cm i.d. × 15 cm, 30 ml/ hr × 8hr/d × 3 d 1 ppm; TDTMA Br (10 × cmc)	Maximum elution at 3.1 relative pore volume with tailing for ai in breakthrough curve. TDTMA Br significantly reduced leaching.	Sánchez et al. (2003a)
41	Diuron 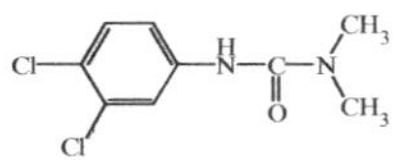	Sandy clay loam, 2 in i.d. × 20 in, 5 in/hr × 24–48 hr 10 lb ai/acre; 4 anionic, 6 cationic & 13 non-ionic surfactants (1 & 10 w/v%)	Cationic surfactants (dilauryl or dicoco dimethyl ammonium chloride) greatly reduced the leaching depth.	Bayer (1967)
		Sandy loam, 3.6 cm i.d. × 10.7 cm, 0.8 cm/hr × 3 hr 5 ppm; Soil penetrant 3685® (alkyl polyoxyethylene ethanol, 0–10,000 ppm)	Reduced (2,000 ppm) & greatly increased (10,000 ppm) leaching by penetrant.	Huggenberger et al. (1973)

(continued)

Table 9 (continued)

No.	Pesticide/structure	Experimental conditions[a] Rate[b], surfactant	Effects on mobility	Reference
42	Linuron 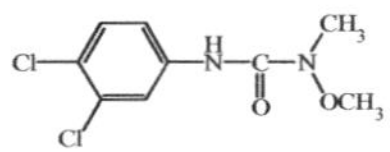	Loamy sand, 5 cm i.d. × 40 cm, 60 ml/d × 33 d 1000 ppm; TDTMA Br & SDS (2 & 15 g/cm^2)	Retardation factor in leaching was >> 9.5 (+ TDTMA Br), 9.04 (unamended), 2.75 (+SDS)	Sánchez-Camazano et al. (2000a)
53	Alachlor 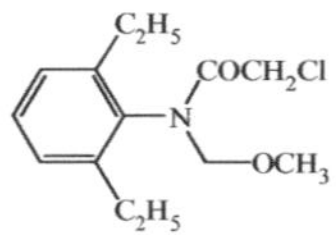	Sandy, 5 cm^2 × 25 cm, 500 m^3/ha for 3–5 hr 2 kg ai/ha Alanex® EC; Organo-clay (Na montmorillonite Swy-1+ 0.5 or 0.8 mmol surfactant/g clay) formulation containing 13% ai	EC leached to the bottom of column and HDTMA-clay slightly reduced leaching. Benzyltrimethylammonium–clay gave maximum leaching depth of 10 cm.	El-Nahhal et al. (1998)
54	Triallate 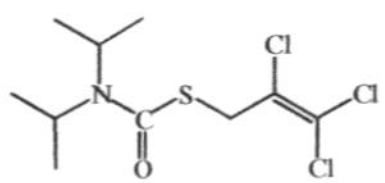	Sily clay loam, 2.2 cm i.d. × 15 cm, 2.54 cm/hr × 6 hr 4.48 kg/ha granule; Emulsifier (nonionic : anionic = 5:1, v/v; 3-fold amount to ai)	Increased leaching from 5 cm to 10 cm but not eluted.	Beestman and Deming (1976)
55	Metolachlor 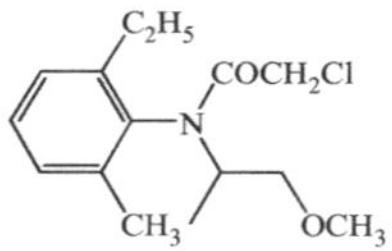	Farm field, 5.9 cm i.d. × 30 cm, 250 ml in total NA; Commercial EC, Organo-clay (Na montmorillonite, + 50% CEC equiv phenyltrimethylammonium (PTMA) or 100% CEC equiv HDTMA) formulation containing 1 & 2 % ai	Organo-clay formulation retarded leaching. Mobility; PTMA-1% ai < PTMA-2% ai < HDTMA-1% ai < HDTMA-2% ai < EC	Singh (2006)

56	Sulfentrazone 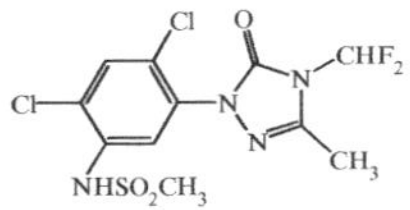	Sandy loam, 100 cm^2 × 20 cm, 50 ml/10 min × 10 0.7 kg ai/ha; Boral® WDG, organo-clay (Na mont-morillonite Swy-2 + 2.5–10 mM octa-decyltrimethyl-ammonium (ODTMA) + ai) formulation	Organo-clay greatly reduced leaching.	Polubesova et al. (2003)
57	Dicamba 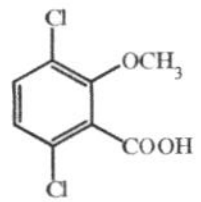	Sandy loam, 4.4 cm i.d. × 50 cm, equiv to 5–20 cm rainfall 8.96 kg ai/ha; Tween 80 (10^3–10^5 ppm)	Enhanced leaching at all concentrations tested with detection of dicamba in leachates.	Foy (1992)
15	Trifluralin 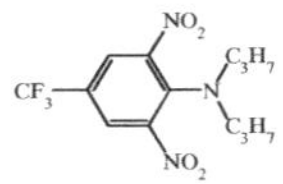	Sandy, 5 cm i.d. × 35 cm, 30 ml per 48 hr 45 kg ai/ha; Tronic®, Wetting agent Ciba®, Triton X-100 (2.0 v/v%)	All surfactants increased downward movement especially when soil was preliminarily wetted	Koren (1972)
58	Bromacil 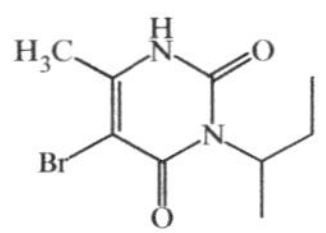	Fine sand, 7 cm i.d. × 8 cm, 94 ml/hr (4 pore volume) 5 kg ai/ha, Hyrar X® 80 WP; 3 ethoxylated amines, 4 quaternary ammoniums (Adogen®), 4 substituted imidazolines (Monazoline®) (weight equiv to ai)	≥ 90% of the applied dose was eluted after addition of 2 pore-volume water. No effect by all surfacrtants on leaching.	Tan and Singh (1995)
13	Atrazine 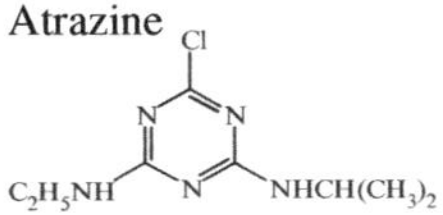	Sandy loam, 3.6 cm i.d. × 10.7 cm, 0.8 cm/hr × 3 hr 5 ppm ai; Soil penetrant 3685® (alkyl polyoxyethylene ethanol, 0–10,000 ppm)	Insignificant effect on leaching. Increased adsorption to soil at ≤ cmc(2,000 ppm) but decreased one at >cmc	Huggenberger et al. (1973)
		Sandy loam, 4.4 cm i.d. × 50 cm, equiv to 5–20 cm rainfall 8.96 kg ai/ha; Tween 80 (10^3–10^5 ppm)	Enhanced leaching at all concentrations tested with detection of atrazine in leachates.	Foy (1992)

(continued)

Table 9 (continued)

No.	Pesticide/structure	Experimental conditions[a] Rate[b], surfactant	Effects on mobility	Reference
		Sandy loam & loam, 5 cm i.d. × 25 cm, saturated flow 3 (sandy loam) & 1.5 (loam) kg ai/ha; Sulphonic®, Rexol® & Rexonic® (200–3000 ppm)	Anionic surfactant (Sulphonic®) greatly reduced leaching.	Abu-Zreig et al. (2000)
46	Norflurazon 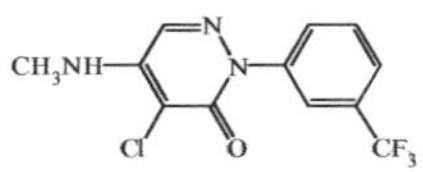	Fine sand, 7 cm i.d. × 8 cm, 94 ml/hr (4 pore volume) 5 kg ai/ha, Solicam® 80 DF; 4 quaternary ammoniums (Adogen®), 4 substituted imidazolines (Monazoline®) (weight equiv to ai)	No effect of Adogen® on leaching but Monazoline® reduced leaching to 60–80% of the control.	Tan and Singh (1995) Singh et al. (2002b)
32	Imidacloprid 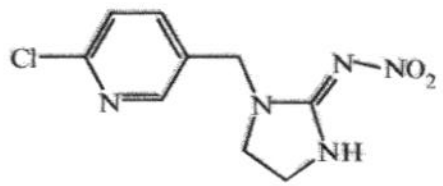	Sandy loam, 7 cm i.d. × 25 cm, 1 mL/min until collection of 0.5 L × 5 leachates. 100 ppm ai; Gaucho® 70 WS, Confidor® 200 SL, Admire® 350 SC	More leaching of formulation than ai with the order of WS > SL > SC.	Gupta et al. (2002)
59	Hexazinone 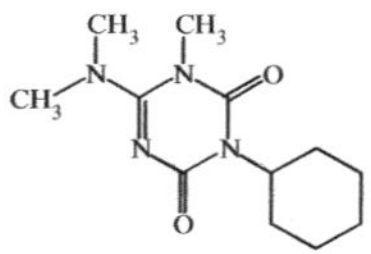	Sandy clay, 5 cm i.d. × 30 cm, 25 ml/d × 60 d 10 kg ai/ha; Velpar® WP, organo-clay (montmorillonite SAz-1 + 100% CEC equiv HDTMA + 4.4 w/w% WP) formulations	Maximum in breakthrough curve delayed by 4 d than WP with less leaching	Celis et al. (2002)

N.A., not available.
[a]Soil properties, column size, flow rate of water and duration.
[b]Application rate of pesticide.

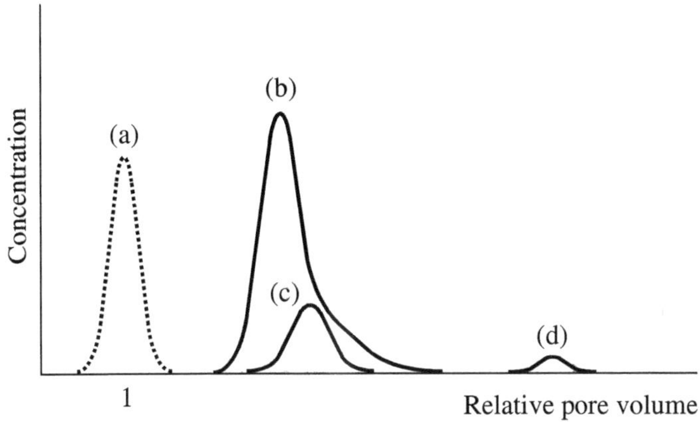

Fig. 9 Breakthrough curves of leachates: (a) tracer such as chloride ion, (b) pesticide only, (c) and (d) pesticide in formulation

The physicochemical properties of pesticide are also an important factor for surfactant effects on its mobility, which was confirmed by the leaching behavior of four herbicides in the presence of Tween 80 (Foy 1992). Nonionic surfactants are considered to increase water penetration by increased dispersion of soil. However, by comparing their effects on the penetration depth of water with those on the mobility of dinitroaniline herbicides, Koren (1972) concluded greater importance of surfactant-controlled adsorption–desorption balance than water movement. Singh et al. (2002a) also demonstrated insignificant effects of 12 commercial adjuvants including organosilicones on downward water penetration in a soil column. The effect of cationic surfactants has been investigated in column leaching studies of norflurazon (46) (Singh et al. 2002b; Tan and Singh 1995). Addition of quaternary ammonium Adogen surfactants to dry flowable formulation of (46) at 1:1 (w/w) ratio slightly reduced mobility at an early stage of leaching, but the substituted imidazoline Monazalin with an optimized mixing ratio significantly reduced the leaching.

The effect of a cationic surfactant has been more clearly demonstrated in a column leaching study of linuron (42) (Sánchez-Camazano et al. 2000a). Amendment of soil by HDTMA Br at the rates of 2 and 15 ton/ha delayed the peak in BTC at a flow rate of 60 mL/day from 3–4 PV to 8.5–>9.5 PV with much smaller eluted amounts. Under the same conditions, anionic SDS reduced retention in the column with the peak in BTC at less than 3 PV due to solubilization of (42) to micelles. A similar soil amendment with 4 w/w% TDTMA Br markedly increased retention of methidathion (53), as evidenced by the significant delay of the BTC peak from 3 PV without amendment to nearly 100 PV (Sánchez et al. 2003a). Lichtenstein et al. (1967a) reported increased mobility of parathion (3) and aldrin (52) in a soil column when 1% anionic SDS was used for elution. Field lysimeter study of atrazine (13) has shown that application with a 10-fold amount of anionic SDBS on a weight basis causes increases of its mobility in soil and the eluted amount (Scheunert and Korte 1985). In contrast, the anionic surfactant Sulphonic greatly reduced downward movement of atrazine (13) (Abu-Zreig et al. 2000). Radioactive residues in leachates decreased from 75%–78% to 13%–52% of the applied ^{14}C by addition of Sulphonic, but nonionic Rexol and Rexonic slightly increased the ^{14}C residues to 80%–87%.

When a pesticide formulation is used for a column leaching study, it seems difficult to predict pesticide leachability. Addition of emulsifier consisting of nonionic and anionic surfactants (5:1, w/w) to granules of triallate (54) slightly increased its movement (Beestman and Deming 1976). Differences in pesticide movement between formulations have been examined for atrazine (13), imidacloprid (32), and metolachlor (55), and higher movement in WP and EC was observed (Bowman 1993; Gupta et al. 2002). Based on accumulated evidence, the combination of pesticide and the type and concentration of surfactant make leaching behavior very complex.

To avoid the undesirable contamination of groundwater, organoclay formulations improving the mobility of pesticides have been developed. El-Nahhal et al. (1998) prepared the organoclay formulation of alachlor (53) by using montmorillonite containing $HDTMA^+$ and benzyltrimethyl-ammonium ($BTMA^+$) ions equivalent to the clay CEC. Although the $HDTMA^+$-organoclay slightly reduced leaching of (53) in laboratory columns where the commercial EC formulation caused its leakage from the 25-cm column, the usage of $BTMA^+$ resulted in a maximum leaching depth of 10 cm. The latter organoclay formulation demonstrated, in the field leaching study, limited movement of (53) to depth of 7 cm, whereas distribution at 12–15 cm was observed for the EC. A similar reduced mobility in an organoclay formulation compared with a commercial EC has been reported for metolachlor (55) by using 50% CEC equivalent $PTMA^+$, but the effect with $HDTMA^+$ was minimal (Singh 2006). Celis et al. (2002) demonstrated the usefulness of $HDTMA^+$-modified montmorilonite to reduce the leaching of hexazinone (59). In contrast to no retention of (59), the peak in BTC for its $HDTMA^+$-clay formulation was observed at 2.5 PV, corresponding to a delay of 4 d with less elution. A similar effect of $HDTMA^+$ was reported for 2,4-D (26), and the longer mixing period to incorporate (26) to the organoclay with a larger volume of an organic solvent was found to be important in reducing leaching of (26) (Hermosín et al. 2006). The very low retention of rimsulfuron (29), together with its metabolites, in a soil column has been reported for its commercial formulation (Martins and Mermoud 1999). $ODTMA^+$-modified organoclay formulations have been developed for sulfosulfuron (12) and sulfentrazone (56) and succeeded in reducing their downward movement more than the commercial ones (Mishael et al. 2003; Polubesova et al. 2003).

C Biodegradation

Both enhancement and inhibition of microbial degradation of organic chemicals by synthetic surfactants and biosurfactants have been reported (Makkar and Rockne 2003; Miller 1995; Rouse et al. 1994). More bioavailability of a chemical via emulsification, solubilization, and partitioning by surfactant monomers and micelles accounts for the enhancement, while the direct interaction of microbial cells with these would modify the function of cell membrane and enzymes, leading to the inhibition. Allen et al. (1999) reported the inhibition of PAH biodegradation by Triton X-100 would originate from some disruption of the cell membrane of microbes as no change of the dioxygenase activity was detected *in vitro*. Effects of surfactants are known to be highly dependent not only on the physicochemical properties such as MSR relating to their chemical structures but also concentration below or above cmc. As an example of the former factor, the more hydrophobic Tween surfactants made the biodegradation of phenanthrene by *Mycobacterium* sp. more rapid (Guerin and Jones 1988). There are several possible processes in uptake of a chemical by microbes (Fig. 10). A water-soluble chemical can be directly taken up [step (a), Fig. 10] but a less soluble chemical would be solubilized by a surfactant monomer or micelle, followed by the transfer to microbial cells [steps (e) and (f)], as well as their modification of the cell membrane and the enhanced attachment of cells to solid surfaces [steps (b) and (c)]. In the presence of soil, the effect of surfactant on soil structure and adsorption/desorption process should also be considered (Tiehm et al. 1997).

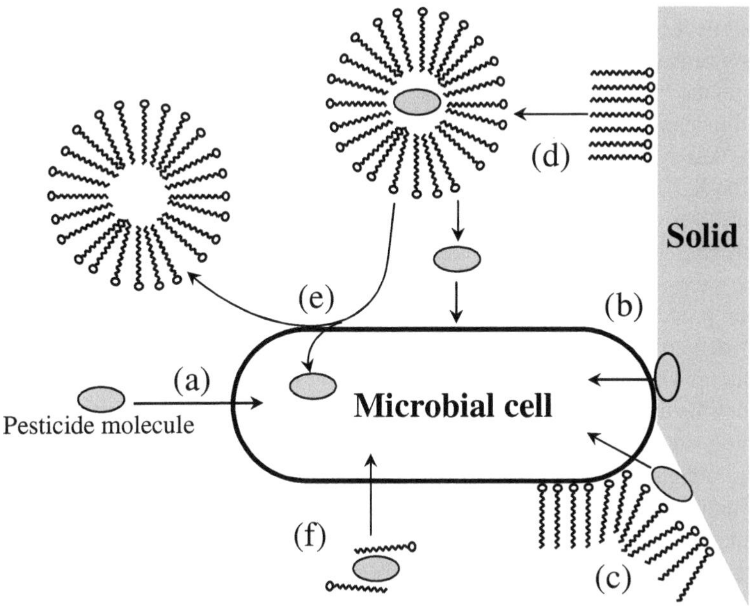

Fig. 10 Possible mechanism on microbial uptake of pesticide. (a) Direct uptake of dissolved pesticide. (b) Uptake via direct attachment of cell on solid phase of pesticide. (c) Uptake via surfactant-induced attachment of cell. (d) Increased mass transfer of pesticide to aqueous phase. (e) Uptake of solubilized pesticide from micelle. (f) Uptake via nonmicellar enhancement of solubilization

Simple Organic Chemicals

The mechanism of the surfactant-modified biodegradation of a chemical has been investigated extensively on PAHs and long-chain alkanes by using isolated microbes, microbial consortia, and intact soils. Laha and Luthy (1992) reported that several types of nonionic surfactants at concentrations above cmc inhibit the bacterial mineralization of phenanthrene, and their presence at lower concentrations shows an insignificant effect or delayed mineralization. They proposed the reduction of microbial enzymatic activity or less transport of a chemical from micelles to the cells as possible reasons (Laha and Luthy 1991). Inhibition of bacterial growth with an effect on enzymatic activity by nonionic surfactants was reported for dibutyl phthalate-degrading soil bacteria (Chao and Lin 2006). The promoted mineralization of phenanthrene and biphenyl at lower concentrations of nonionic surfactants than cmc has been shown by Aronstein et al. (1991), and their enhanced desorption from soil was considered to increase bioavailability to microbes. Increased solubilization by surfactants could facilitate the mineralization of decane by the two gram-negative bacterial strains, even in the presence of micelles with the biodegradation rate following the Monod equation (Bury and Miller 1993). Zhang and Miller (1994) examined the effects of rhamnolipid on the biodegradation of octadecane by four isolated microorganisms producing the biosurfactant and showed that promotion and inhibition of biodegradation are highly dependent on both surfactant concentration and bacterial species.

The effect of surfactant on biodegradation has been also theoretically investigated by many researchers. For less water-soluble chemicals such as PAHs, the dissolution rate from their solid states becomes a rate-determining step of microbial degradation. In the biodegradation of solid phenanthrene and naphthalene by two *Pseudomonas* strains, Volkering et al. (1995) reported the enhanced dissolution of PAHs above cmc by Triton X-100 and Brij 35 together with more biodegradation, but bioavailability of the micellized PAHs to the microbes was considered less than those dissolved in an aqueous phase (see Fig. 10d). Similar results were reported by Grimberg et al. (1996), and the excess amount of surfactants forming more micelles was found to reduce bacterial growth by reducing bioavailable phenanthrene. Furthermore, Mulder et al. (1998) demonstrated through the enhanced biodegradation of naphthalene in the presence of six nonionic surfactants that a mechanistic mass-transfer model well describes the dissolution of solid naphthalene by the surfactants and that the biomass formation rate by a *Pseudomonas* strain increases concomitantly with the mass-transfer rate under naphthalene dissolution-limited conditions.

Assuming that the transfer of a chemical between phases is instantaneous relative to biodegradation and the surfactant does not alter the specific activity of biomass, Guha and Jaffé (1996a,b) have kinetically analyzed the micellar effect of nonionic surfactants on the biodegradation of ^{14}C-phenanthrene by a mixed enrichment culture isolated from a petroleum-contaminated soil. They separated step (e) in Fig. 10 into the mass transfer of the filled micelle to a cell surface, the attachment of the filled micelle to the cell surface as a hemimicelle, and the transfer of a chemical from the hemimicelle to the cell. The second process was defined as the ratio of the micellar diffusion relaxation time and the relaxation time for the micelle formation/breakdown. The third process was assumed to be proportional to the gradient of the chemical concentration across the cell. The factor "f," a bioavailable fraction of the micellized chemical, was then expressed as a function of the concentration of micellized surfactant. By analyzing the change of biomass and production of carbon dioxide as a function of a surfactant concentration and time, they have succeeded in reproducing the bioavailability of the micellized phenanthrene experimentally observed. The f values of Trirton X-100 and Brij 30 decrease from 0.8 with an increase of the surfactant concentration finally to zero, but that of Brij 35 was kept close to zero at all concentrations tested. The insignificant bioavailability from the Brij 35 micelles may originate from hydrophilicity of its long polyethoxy chain not well interacting with hydrophobic cell surface. Brown (2007) introduced the detailed process of a hemimicellar formation on the cell surface and clarified that the surfactant-dependent mass transfer of a hydrophoboic chemical into a cell originates from the variation of the surfactant sorption on the cell surface. The unavailability of the intercalated PAHs in $HDTMA^{+}$- and $TDTMA^{+}$-modified clays to microbes showed the importance of the release of a chemical from the partitioned medium consisting of surfactants (Crocker et al. 1995, Theng et al. 2001).

Pesticides

The effects of adjuvants in formulation on the biodegradation of pesticide in soil or sediment under aerobic and anaerobic conditions are summarized in Table 10. The dominant controlling factors on the effects are the type of surfactant and its concentration, but the complex composition in pesticide formulation usually makes it very difficult to predict the effect of adjuvants on the biodegradation of pesticide.

Amonette and O'Connor (1990) have shown the slower initial degradation of 2,4-D (26) in soils treated with alkylphenoxy ethoxylate having a lower HLB value, which was accounted for by less bioavailability of (26) due to more solubilization to micelles. The anaerobic biodegradation of DDT (1) by anaerobes was enhanced by addition of approximately 10-fold amounts of Brij-35 or Triton X-114, with the initial reaction rate increasing by a factor of 1.5–3 (You et al. 1996). The main degradate was DDD via reductive dechlorination of (1), and the presence of the surfactants did not affect the ratio of non-DDD products. Walters and Aitken (2001) reported increased anaerobic biodegradation of (1) in soil slurries by addition of Brij 30 in concomitant with the increased solubilization of (1). Insignificant enhancement by Tween 80 at the concentrations of 5-to 20 fold cmc was found for the anaerobic degradation of lindane (34), but its β- and δ-isomers were biodegraded slightly more by addition of the surfactant at 20-fold cmc (Quintero et al. 2005). In contrast, Triton X-100 at the concentration of 5-fold cmc remarkably inhibited its biodegradation. The addition of nonionic micellar surfactant increased the amount of mineralization in the soil metabolism of triticonazole (48) formulation, while the higher concentration of alkylnaphthalene sulfate surfactant caused an inhibitory effect (Beigel et al. 1999; Charnay et al. 2000).

Table 10 Effect of surfactant on biodegradation of pesticides in soil

No.	Pesticide/structure	Soil[a]/surfactant[b]	DT_{50} or MR[c]	Reference
34	Lindane 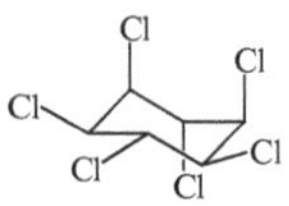	Sandy loam soil, anaerobic (100, 9.8, NA, 4.04, 30) + Triton X-100 (5 × cmc) + Tween 80 (5 × cmc)	1.0 d 3.1 d 1.1 d	Quintero et al. (2005)
3	Parathion 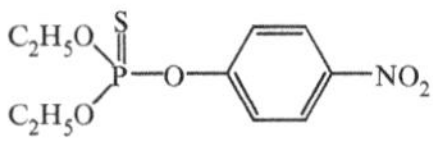	Silt loam soil, aerobic (10, NA, NA, NA, 30) + ABS (1% w/w to soil)	4.8 d 15.9 wk	Lichtenstein (1966)
38	Diazinon 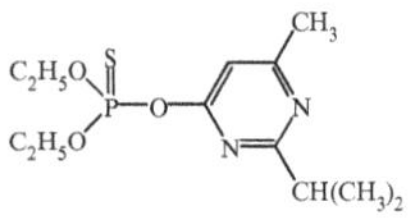	Silt loam soil, aerobic (10, NA, NA, NA, 30) + ABS (1% w/w to soil)	1.3 wk 3.6 wk	Lichtenstein (1966)
39	Azinphos-methyl 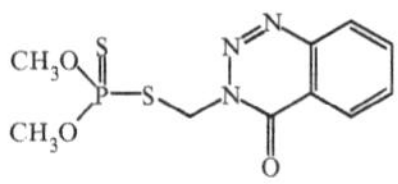	Silt loam soil, aerobic (20, NA, NA, NA, 30) + EC formulation (1/600, 2 lb/gal)	6 d[d] 22 d[d]	Schulz et al. (1970)
40	Chlorotoluron 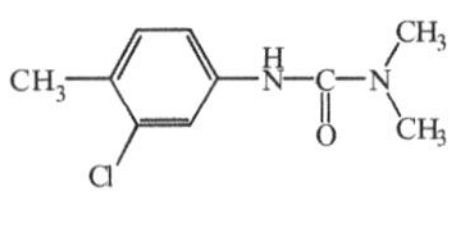	UK soil (60% water-holding capacity), aerobic, (10, 2.76, 73/11/16, 7, 20) + 50% w/v suspendable concentrate + 80% w/v wettable powder	NA 80 d 81 d	Cotterill (1988)
42	Linuron 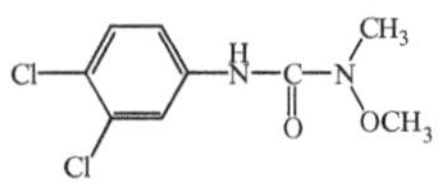	Sandy loam soil, aerobic, nonsterile (60, 0.67, 64.0/15.5/18.1, 7.5, 28±2) + SDS (1.5 g/m^2) + HDTMA Br (1.5 g/m^2)	248 d 147 d 234 d	Rodriguez-Cruz et al. (2001)
		Sterile + SDS (1.5 g/m^2) + HDTMA Br (1.5 g/m^2)	277 d 147 d 204 d	
60	Aldicarb 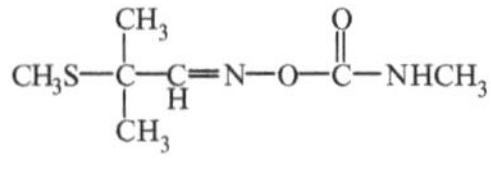	Farmland surface soil, aerobic (20, 2.47, 76/11/13, 7.93, 25) + SDBS (100 ppm) (1000 ppm)	6.13 d 5.06 d 3.92 d	Liu et al. (2003)
61	Benomyl 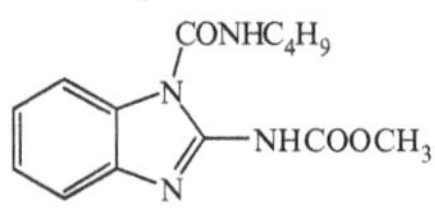	Sandy loam, aerobic (60–97, 3.7, NA/NA/16.4, 7.9, 23) + Dry flowable formulation + Soluble powder formulation	NA 34–92 d 55 d	Sassman et al. (2004)

(continued)

Table 10 (continued)

No.	Pesticide/structure	Soil[a]/surfactant[b]	DT_{50} or MR[c]	Reference
45	Metalaxyl 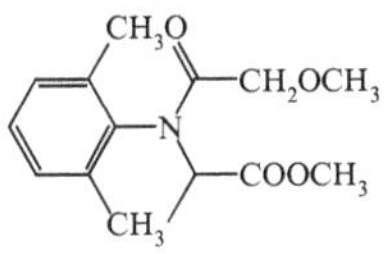	Sandy clay loam, aerobic (1.6, 5.2, 45/24/31, 4.2, 20±2)	18 d	Monkiedje et al. (2003)
		+ EC formulation	18 d	
44	Acetochlor 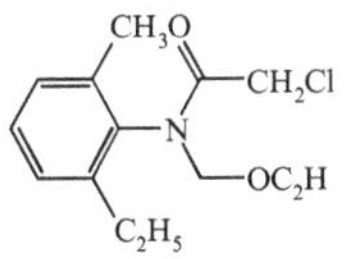	Agricultural field soil, aerobic (10, 2.59, 38/59/3, 8.4, 35±1)	4.6 d	Ye (2003)
		+ SDBS (NA)	6.3 d	
48	Triticinazole 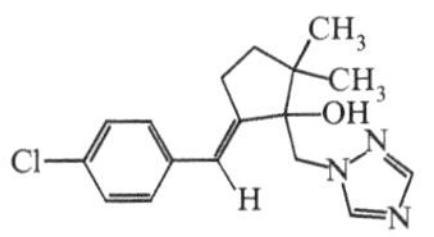	Agricultural field soil, aerobic (4, 17.9, 14.5/54/29.1, 8.2, 28)	1.3×10^{-3} d^{-1}	Beigel et al. (1999)
		+ REAL® blank formulation (NA)	1.6×10^{-3} d^{-1}	
		+ REAL® / Soprophor FLK (NA)	1.6×10^{-3} d^{-1}	
		Loamy clay soil, aerobic (10^3, 1.79, 14.7/55.5/29.8, 8.2, 28±1)	0.21 d^{-1}	Charnay et al. (2000)
		+ Alkyl naphthalenesulfate (1.7 g/kg)	0.08 d^{-1}	
		+ Phosphate tristyryl phenolethoxylate	0.21 d^{-1}	
		+ nonionic (ethylene oxide, fatty acid)	0.05 d^{-1}	
32	Imidacloprid 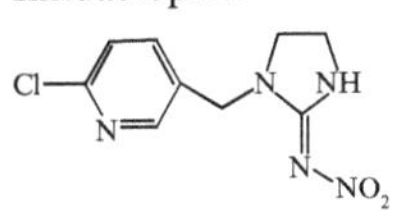	3 soils, aerobic (0.25–0.5, 11–38/12–36/17–53, 5.2–8.5, 28±1)		Sarkar et al. (2001)
		+ Confidor® 200 SL	29–48 d	
		+ Gaucho® 70 WS	36–46 d	

[a]The values in the parentheses are pesticide concentration applied to soil (ppm), organic matter content (%), fraction% of sand, silt and clay, soil pH and incubation temperature (°C).
[b]The value in the parentheses is the concentration or amount of surfactant being applied. cmc, critical micelle concentration.
[c]DT_{50}, degradation half-life (time). MR, mineralization rate ($time^{-1}$).
[d]Time for 95% degradation. NA, not available.

Lichtenstein (1966) reported the slower degradation of organophosphorus insecticides in soils treated with anionic alkylbenzene sulfonate and LAS. Because the microbial population increased by their addition, less bioavailability of the pesticides associated with micelles being more tightly adsorbed on soil seemed to account for this retardation. A similar mechanism was considered to operate for the soil metabolism of acetochlor (44) in the presence of SDBS (Ye 2003). In the case of soil metabolism of linuron (42), SDS was found to enhance its biodegradation but HDTMA Br showed an insignificant effect (Rodriguez-Cruz et al. 2001). SDBS enhanced the aerobic biodegradation of aldicarb (60) in soil by 28%–57%, which was explained by the increased solubility of (60) and the change of soil pH by the surfactant (Liu et al. 2003).

Enhanced aerobic biodegradation even at a higher concentration where synthetic surfactant shows an inhibitory effect has been reported for biosurfactants. Surfactin, a biosurfactant produced by *Bacillus sabtilis*, has been reported to promote the aerobic biodegradation of endosulfan (6,7,8,9,10,10-hexachloro-1,5,5*a*,6,9,9*a*-hexahydro-6,9-methano-2,4,3-benzodioxathiepin 3-oxide) by approximately 30% for both α- and β-isomers without accumulation of the corresponding diols and sulfates (Awasthi et al. 1999). Mata-Sandoval et al. (2000, 2001) investigated the effect of rhamnolipid mixtures produced by *Pseudomonas aeruginasa* UG 2 on the aerobic degradation of coumaphos (7), atrazine (13), and trifluralin (15) by *Streptomyces* PS 1/5 in a liquid medium and soil slurries and showed its effectiveness even at higher concentrations.

Adjuvants in formulation are known to affect the biodegradation of pesticides. Soil metabolism of azinphos-methyl (39) in its EC formulation was retarded by a factor of 3–4 compared with the application of a.i., and much slower degradation was observed for its granule formulation (Schulz et al. 1970). Microbial degradation of some pyrethroids in a culture medium was slightly reduced for their formulations (Grant 2001). In contrast, the effect of formulation on the biodegradation of propetamphos ((*RS*)-[(*E*)-*O*-2-isopropoxycarbonyl-1-methylvinyl *O*-methyl ethylphosphoramidothioate]) in sediments (García-Ortega et al. 2006) and imidacloprid (32) (Sarkar et al. 2001) and chlorotoluron (40) (Cotterill 1988) in soil was insignificant. Monkiedje et al. (2003) reported the insignificant effect of adjuvants in the EC formulation of metalaxyl (45). Bromilow et al. (2003) reported some enhanced biodegradation of several pesticides in water–sediment systems, possibly resulting from the increased microbial population, by the aid of coformulants. The bioavailability of pesticide in organoclay formulation prepared by mixing clay with a quaternary ammonium ion has been reported to be lower than that dissolved in a medium for reasons of inaccessibility of microbes to the intercalated pesticides (Hermosín et al. 2006; Singh et al. 2003).

VII Effects on Behavior in Plants

The application of a pesticide formulation in the field is generally classified into foliar and soil treatments. The behavior of pesticide in the former treatment has been investigated extensively, not only experimentally but also theoretically by many researchers, but fewer investigations are available for the latter. The foliar uptake is initiated by penetration of pesticide through cuticle and sometimes stomata from the formulation deposits, followed by apoplastic, synplastic transport and phloem translocation, whereas root hairs play the most important part in absorption of pesticide from soil (Field and Dastgheib 1996). The adjuvants in formulation including surfactant are considered to affect these processes. When the extent of pesticide uptake is affected, its metabolism in plants may be modified by surfactant (Baloch 2000). However, plant metabolism study is usually conducted by using the solution of a water-miscible organic solvent or the formulation using ^{14}C-labeled pesticide in accordance with good agricultural practice (OECD 2007), and hence relevant information on the surfactant effect is very limited.

A Root Uptake

The uptake of pesticide by roots is considered to be initiated by its absorption and translocation through root hairs. Briggs et al. (1982, 1983) introduced the concept of transpiration stream concentration factor (TSCF), defining the ratio of pesticide concentration in the transpiration stream and an aqueous phase outside the root, which is a function of log P for nonionic chemicals. The root concentration factor (RCF) expressing a similar ratio between the root and the aqueous phase was found to be also proportional to log P. Because surfactants can solubilize a chemical by association with its monomers and micelles, they are considered to affect these partition processes. Li et al. (2001) examined the effect of Tween 80 on the distribution of ^{14}C-phenanthrene applied as an aqueous solution in closed systems including wheat seedlings and lava. The surfactant slightly reduced the ^{14}C-distribution in the root, probably because of more solubilization of PAH in the aqueous phase containing lava. Jiang et al. (2005) reported a similar effect by LAS on phenanthrene in the same system using rushes instead. The distribution of ^{14}C in shoots was around 50% of the applied ^{14}C in wheat but only about 2% for rushes. In both cases, more mineralization with formation of polar metabolites was observed, which was caused by the enhanced microbial degradation in the aqueous phase. Although 10%–40% of the ^{14}C-labeled dioxin congeners in the 0.05% Tween 80 solution was absorbed by the roots of soybean and oat seedlings, insignificant translocation of ^{14}C to their tops (<1%) was observed (Isensee and Jones 1971).

Enhanced root uptake of PAHs by 6%–35% from aqueous solution of Brij 35 at concentration less than cmc has been reported for ryegrass seedlings (Gao et al. 2006). The excess amount of the surfactant reduced the uptake by its phytotoxicity and less bioavailability caused by greater solubilization of PAHs to micelles in the aqueous phase. In the case of pesticids, information on the surfactant effect on root uptake is very limited. Lichtenstein et al. (1967b) reported the effect of 0.005% LAS on the root uptake of the several pesticides by pea seedlings. LAS greatly reduced the root uptake of parathion (3) but gave insignificant effects for lindane (34), diazinon (38) and aldrin (52). The translocation of (34) was reduced to one-seventh of that in the absence of LAS but more (52) with formation of dieldrin (62) was detected in the green part by addition of LAS. The increased efficacy of benomyl (61) in its soil application with Tween 20 for cotton may imply enhanced root uptake (Rawlins and Booth 1968).

Application of SDBS with about a 10-fold amount of atrazine (13) in an outdoor lysimeter study has shown that root uptake of ^{14}C and (13) in maize after the first season increases by factors of 2 and 3, respectively (Scheunert and Korte 1985). However, in field residue trials of (13), not only preemergence application of the flowable and wettable powder formulations but also postemergence application of (13) mixed with three different adjuvants gave insignificant differences in residues of the metabolites in oats (Khan et al. 1981). Butachlor (74) applied to soil was absorbed and translocated by rice plants, but its dissipation rate was independent of formulation type (Kulshrestha 1987). These limited studies may show possible enhancement or inhibition of the root uptake of chemicals including pesticides by surfactant for some simple model systems, but under field conditions the effects are likely to be diminished by various factors such as dilution, adsorption to soil, and bacterial degradation of surfactant.

B Foliar Uptake

Adjuvants in formulations, especially surfactants, are known to increase the foliar penetration of pesticide (Bentson 1990; Bukovac et al. 2003). The surfactant having a higher HLB value tends to increase hydration of the cuticle and as a result to enhance the permeance of a hydrophilic chemical (log $P < 3$), whereas that having a lower HLB is considered to cause the enhanced permeance of a hydrophobic chemical (log $P > 3$) via an increase in the fluidity of cuticular waxes (Baur 1998; Hess and Foy 2000; Wang and Liu 2007). Penetration of pesticide and surfactant through the wax region and the underlying cutin layer has been studied extensively by either adsorption and desorption on a wax-coated disk (Schreiber and Schönherr 1993) or mass transfer through an enzymatically isolated cuticular membrane in a diffusion cell (Bauer and Schönherr 1992; Schönherr and Riederer 1989). It is considered through diffusion experiments that the surfactant above cmc depresses the partition of pesticide to a cuticle membrane by its solubilization to micelles and that the permeance of the pesticide in cuticle increases when the surfactant is concomitantly dissolved into the cuticle (Schönherr et al. 1991). The copermeation of pesticide and surfactant molecules has been shown by Schreiber (1994) through the desorption of pentachlorophenol from reconstituted barley waxes in the presence of dodecyl octaethoxylate surfactant. Burghardt et al. (1998, 2006) found that the extent of enhanced diffusion of 15 monodisperse alcohol ethoxylates is proportional to their concentration in the isolated cuticle with a concomitant increase of pesticide diffusion. The presence of surfactant was found to increase the free volume available for a pesticide diffusion, showing the plasticizing effect of surfactant.

Stevens et al. (1988) examined the effect of nonylphenol octaethoxylate on the foliar absorption and translocation of ^{14}C-glucose and 14 pesticides for four plant species and shown that the log (% uptake) and log (% translocation) have a parabolic dependence on the log P value of pesticide with a maximum at 0 and 1.5–2.2, respectively. Through the foliar uptake of ^{14}C-labeled methylglucose and 4 pesticides having the log P values of −3.0 to 6.5 by wheat and field bean, comparison of uptake in the presence of four alcohol ethoxylates having the HLB value of 11 to 16 demonstrated that more hydrophilic surfactant (higher HLB) enhances the uptake of more hydrophilic pesticides ($-3.0 < \log P < 0.8$), but the hydrophobic compounds ($2.1 < \log P < 6.5$) are taken up more in the presence of the surfactant with a lower HLB value (Stock et al. 1993). This classification was in agreement with the results of uptake studies (Coret and Chamel 1995; Holloway et al. 1992).

In the case of an ionizable pesticide such as sulfonylurea herbicide, pH of the medium controls its log P value, and then the effect of the surfactant HLB value on the foliar uptake has been found to change with pH (Green and Hale 2005). Forster et al. (2004, 2006a,b) revealed a very simple equation to predict the foliar uptake of pesticide in living plants by using several pesticides and various kinds of adjuvants. Adjuvants increased the foliar uptake of pesticide in accordance with the equation "Uptake in mol per unit area" = $a[\mathrm{ID}]^b$ where ID is an initial dose in mol per unit area with a and b being the constants. Uptake in the presence of adjuvants having a higher molecular weight than 1350 deviated from this equation, possibly because of less concomitant penetration of the adjuvant into cuticles. In addition to the usual foliar uptake through the cuticle, the direct infiltration of stomata by a pesticide formulation has been demonstrated to be possible when the surface tension of the surfactant is less than 23 mN/m using organosilicones. Stevens et al. (1992) found 20%–50% infiltration of the applied glyphosate (*N*-(phosphonomethyl)glycine) via the stomata of bean, oat, and wheat by using 0.5% Silwet L-77 surfactant.

The effect of adjuvants including surfactant on foliar uptake has been examined for intact plants, isolated cuticle, and model systems such as reconstituted waxes using the formulation and ^{14}C-labels of many pesticides. For four enzymatically isolated cuticles in the presence and absence of waxes, the effect of SDS and 16 nonionic surfactants on the permeability of 2,4-D (26) was investigated (Schönherr and Bauer 1992). Partition of (26) to the cuticle membrane decreased due to solubilization to micelles, but the penetration of nonionic surfactants having a polyethoxylated chain plasticized the membrane, which resulted in uptake activation. This plasticizing effect was dependent on the plant species and found to be greater for the ester-type surfactants such as diethyl suberate and sebacate than decyl ethoxylate in a mobility experiment with 2,4-dichlorophenoxybutyric acid (Šimáňová et al. 2005). The importance of hydrophilicity in alcohol ethoxylates was reported for the foliar uptake of glyphosate by intact oat and field bean plants and the surfactant having a low HLB value became a competitor for uptake (van Toor et al. 1994).

A similar trend for the uptake of glyphosate was observed for isolated cuticles, and differential scanning calorimetry clearly demonstrated the increased fluidity of the cuticular wax in the presence of nonionic surfactant (Coret and Chamel 1995). Sharma et al. (1976) examined the foliar uptake of ^{14}C-dufenzoquat (1,2-dimethyl-3,5-diphenyl-1*H*-pyrazolium) by wild oat and reported marked enhancement by a factor of 7 in the presence of 0.4% Triton X-100. The foliar uptake of pesticides having an aromatic acid skeleton has been investigated in the presence of surfactants. Anionic surfactants were found effective in enhancing the uptake of picloram (4-amino-3,5,6-trichloropyridine-2-carboxylic acid) by leaf disks of two *Eucalyptus* species, and the amount of leaf wax likely affected the uptake (Sands and Bachelard 1973).

Thompson and Nissen (2000) have explained the moderate tolerance of soybean to carfentrazone-ethyl (ethyl 2-chloro-3-{2-chloro-5-[4-(difluoromethyl)-4,5-dihydro-3-methyl-5-oxo-1*H*-1,2,4-triazol-1-yl]-4-fluoro- phenyl}propionate) partly by the enhanced foliar absorption by nonionic surfactant, indicating the different effects of the surfactant dependent on plant species. The ionization of dicamba (57) was related to the effect of surfactant on the foliar uptake by soybean (Petersen et al. 1985). The tested surfactants slightly affected the uptake of dimethylamine salt of (57) irrespective of ionic type, whereas all the surfactants increased the uptake of the potassium salt by 35%–56%. Furthermore, (57) treated at the first trifoliate was more translocated to the second by a factor of 1.5 in the presence of the surfactants.

Lamoureax and Rusness (1995) found marked enhancement of foliar uptake of ^{14}C-quinclorac (3,7-dichloroquinoline-8-carboxylic acid) by leafy spruce by Lutensol, consisting of an aromatic solvent and C_{10}-Guerbert alcohol surfactant, leading to formation of three glucose conjugates. The nonylphenoxy ethoxylate surfactant, Agral 90, was found to increase not only retention but also foliar uptake and translocation of imazamethabenz (2-[4,5-dihydro-4-methyl-4-(1-methylethyl)-5-oxo-1*H*-imidazol-2-yl]-4 (or 5)-methylbenzoic acid) in wild oats (Smith and Chow 1990). In contrast, the hydrophobic ^{14}C-diflubenzuron (1-(4-chlorophenyl)-3-(2,6-difluorobenzoyl)urea) was scarcely taken up by cotton leaves, with insignificant differences in ^{14}C distribution between aqueous suspension and oil emulsion formulations (Mansager et al. 1979). Foliar uptake of bentazone (28) applied to the adaxial leaf surface of mustard and bean plants related to the chemical structure of the applied nonionic surfactants (Liu 2004). Uptake efficiency by surfactants having a similar polyethoxylate chain decreased when they had the following structures: C_{13}/C_{15} linear alcohol > C_{10} linear alcohol > C_{16}/C_{18} linear alkenols > nonylphenol > octylphenol. For surfactants having the substructure of C_{13}/C_{15} linear alcohol, the lower polyethoxy chain gave the higher uptake. Humidity was also one of the controlling factors enhancing foliar uptake and translocation of pesticide. Smith and Nalewaja (1972) have reported greater than 10-fold enhancement in the uptake of atrazine (13) by yellow foxtail in the presence of 10% phytobland oil containing 1% Triton X-207 surfactant, and its extent was greater under higher humidity.

VIII Effects on Pesticide Residues

Possible effects of adjuvants in formulation on pesticide behavior should be confirmed through the field trials using commercial formulations. However, a field trial is usually conducted by using the most appropriate formulation exhibiting the highest efficacy of the pesticide, and thus the information on side-by-side comparison of the residue profiles is generally very limited, not only between the a.i. and its formulation but also among different formulations. By surveying the accumulated evidence on residue trials of some pesticides, the effects of adjuvants and surfactants are summarized in Table 11.

The method of pesticide application, its interception by plants, and weather conditions are known to be the dominant factors controlling foliar pesticide residues, and the type of formulation has been demonstrated to be mostly a minor factor in the dissipation half-life (DT_{50}) of pesticide (Willis and McDowell 1987). The DT_{50} values for organochlorine (Harrison et al. 1967; Mukherjee and Gopal 1999) and organophosphorus pesticides (Günther 1969; Miller et al. 1969; Montemurro et al. 2002; Pree et al. 1976) decreased in the order of emulsifiable concentrate (EC) > wettable powder (WP) > dust (D) and WP > EC > D, but their differences were less than a few days and considered not decisive. In some cases, the effect of formulation has been clearly shown. The WP formulation of vinclozolin (17) was considered to reduce volatilization loss from the leaf surface of garden pea (Szeto et al. 1989). The addition of Ortho HDD alkylphenoxy ethoxylate to the WP formulation markedly increased the initial penetration of dimethoate (67) in citrus leaves, but with no effect on translocation from peel to pulp and the dissipation rate (Woodham et al. 1974).

Depending on the formulation type and adjuvants, the amount of an initial deposit and a dislodgeable fraction on crops are known to significantly differ (Asano et al. 1984; Günther 1969; Holloway and Western 2003). Among the same formulation types, the composition of adjuvants affected the DT_{50} value as well as the initial residues (Chang-Yen et al. 1983). Sundaram (1986) reported that the volatility of an adjuvant in relation to droplet size of an applied formulation controls both the initial deposits and dislodgeable residues of fenitrothion (5) on conifer trees and that the presence of non- or low-volatile adjuvant tends to increase the DT_{50} value of residues. Kucharski (2003) reported the longer persistency of several herbicides in leaves and root of sugar beet when the corresponding EC formulations were prepared by addition of different emulsifiers. Marshall and Pree (1993) found different shapes of deposits on apple leaves between EC and WP formulations of propargite (71) by scanning electron micrography. The greater distribution in the EC formulation showed higher efficacy, but insignificant differences in either the initial deposit or DT_{50} were observed. The "sticker" adjuvant was shown to increase the initial deposits of pyrethroid insecticides on soybean and cotton but did not significantly alter their persistence and rainfastness (Reeves 1993). Similar effects were reported for the spreader-sticker adjuvant Bivert in the field application of chlorothalonil (63) and chlorpyrifos (65) to a cranberry bog (Putnam et al. 2003). Higher dislodgeable residues of deltamethrin (69) in staked cucumber were detected by using the concentrate suspension formulation rather than the EC (Franco et al. 2005). Additives in formulations were shown to enhance the photodegradation of azadirachtin (25) by a factor of 5 (Caboni et al. 2002). Cao et al. (2005) developed a unique suspo-emulsion formulation including anatase TiO_2 as a photocatalyst to reduce pesticide persistence.

Table 11 Effect of formulation and its additives on plant and soil residues of pesticides

No.	Pesticide/structure	Residue trial/Rate, period/analysis	Formulation/additives	Effects	Reference
Plant residues					
62	Dieldrin 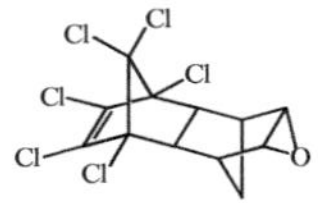	Foliar residues of apple trees 0.02–0.05 ai %, June, 11 wk Hexane extracts, GC	Dieldrex 15® EC 50% Dispersible powder	Initial deposits, DP > EC Persistency, EC > DP	Harrison et al. (1967)
35	Heptachlor 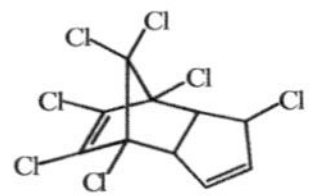	Eureka lemon rind 2 lb ai/100 gal, 350 gal/acre, 60 d NA	30% wettable powder 20% EC	Initial deposits, WP > EC Persistency, EC > WP	Günther (1969)
34	Lindane 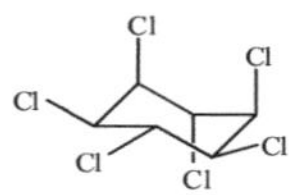	Chickpea pods (pod cover, grains) 400 & 800 g ai/ha, 0–16 d after application and at harvest Partitioned to hexane, GC	1.3% Kanodane® dust Kanodane® 20EC Kanodane® 6.5% WP	Initial residues, WP > D > EC Half-life, WP (4.7–5.1 d) EC (6–7.5 d) D (4.5–5 d)	Mukherjee and Gopal (1999)
63	Chlorothalonil 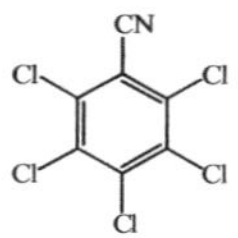	Residues on cranberry 3.5 kg ai/ha, 0–28 d CH_3OH (leaf), hexane-acetone extraction, GC	Bravo® 720 ± Bivert® sticker adjuvant[a] (28 v/v%)	Increased initial dislodgeable foliar and fruit residues with adjuvant	Putnam et al. (2003)
1	DDT 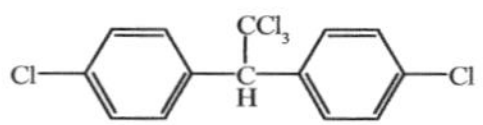	Foliar residues of apple trees 0.1 ai %, June, 11 wk Hexane extracts, GC	Strykol® 25% EC 50% Dispersible powder	Initial deposits, DP ≈ EC Persistency, EC > DP	Harrison et al. (1967)

64	Diclofop-methyl 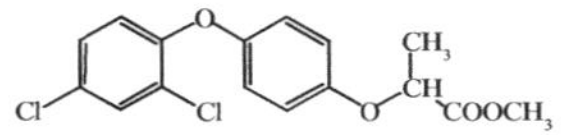	Residues on wheat & field beans 1.135 kg ai/ha, 0–120 hr Acetone extraction, LC-MS[2]	Hoegrass® 37.8% EC + Agral® (1 mL/L), Toli® (7.5 mL/L), Bond® (1.4 mL/L)[b)]	Foliar dissipation, EC+Bond ≥ EC ≈ EC+Agral > EC+Toli	Holloway and Western (2003)
5	Fenitrothion 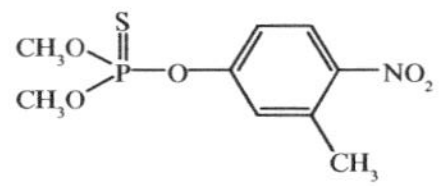	Foliar residues of apple trees 90–23 g ai/ha, 0–29 d $CHCl_3$-CH_3OH rinse & extraction	Accothion® 50% EC Fentro-50® EC	% sub-surface residues, Fentro-50 > Accothion DT_{50} = 0.3–1.4 d (Acc) 0.6–0.9 d (Fen)	Chang-Yen et al. (1983)
		Foliar residues of conifer trees 340 g ai/ha, 0–150 d EtOAc wash & homogenates, GC	4 EC formulations with different diluents	Dislogeable, DT_{50}=13–15 d Penetrated, DT_{50}=16–24 d	Sundaram (1986)
65	Chlorpyrifos 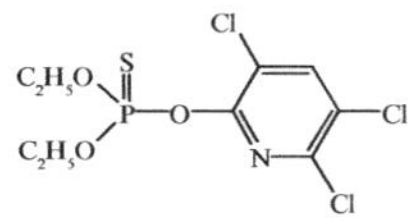	Residues on orange fruits and leaves 680–730 g ai/ha, 0–130 d EtOAc/cyclohexane extracts, GC	Clorpiran® 40 EC Dursban® 75 WG	DT_{50} EC WG Fruits: 24.6 d 42.8 d Leaves: 6.6 d 6.7 d	Montemurro et al. (2002)
		Residues on cranberry 0.8 kg ai/ha, 0–28 d CH_3OH (leaf), hexane-acetone extraction, GC	Lorsban® 4E ± Bivert® sticker adjuvant (25 v/v%)	Increased initial dislodgeable foliar and fruit residues with adjuvant	Putnam et al. (2003)
		Foliar residues on corn 560 g ai/ha, 0–10 d Pentane-CH_2Cl_2 extraction, GC	Lorsban 4E (EC) ai + once-refined soybean oil (Oil)	Initial deposits, oil > EC DT_{50} ~ 1 d (EC, oil)	Wauchope et al. (1991)
66	Phosmet 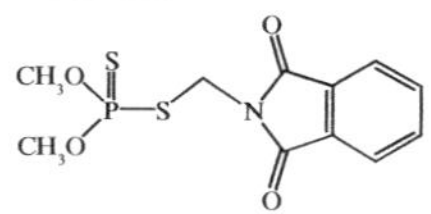	Foliar residues on alfalfa 0.5–1 lb ai/acre, 0–21 d NA, colorimetry	Imidan® 50 WP Imidan® 3 EC	DT_{50} = 1.6–2.1 d (WP), 2.2–2.5 d (EC)	Miller et al. (1969)

(continued)

Table 11 (continued)

No.	Pesticide/structure	Residue trial/Rate, period/analysis	Formulation/additives	Effects	Reference
39	Azinphos-methyl 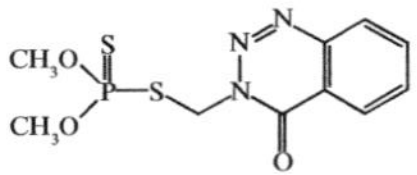	Foliar residues on sugarcane 2 (EC), 15 (G) lg ai/acre, 0 d NA, colorimetry	Attaclay granule (G), EC, Low-volume-concentrate (LV)	Whole plant deposits, EC ≈ LV >> G	Davis et al. (1969)
39	Azinphos-methyl	Foliar residues on apple trees 330 ppm, 0–42 d EtOAc extraction, GC	EC & WP	Initial deposits, EC > WP DT_{50} = 6.3 d (EC) 6.2 d (WP)	Pree et al. (1976)
67	Dimethoate 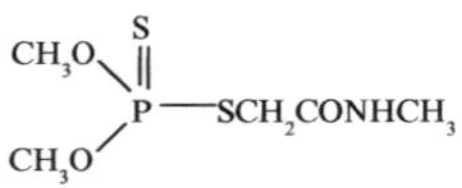	Residues on grapefruits and leaves 2.5 lb ai/acre, 0–14 d CH_2Cl_2 (leaf), acetonitrile (peel), acetone (pulp) extracts, GC	25% WP ± Ortho HDD®c), 25 w/w% to formulation	More rapid penetration of diemthoate into leaves but with similar residues after 2 wk	Woodham et al. (1974)
		Foliar residues on apple trees 330 ppm, 0–42 d EtOAc extraction, GC	EC & WP	Similar initial deposits. DT_{50} = 7.1 d (EC) 5.4 d (WP)	Pree et al. (1976)
68	Aminocarb 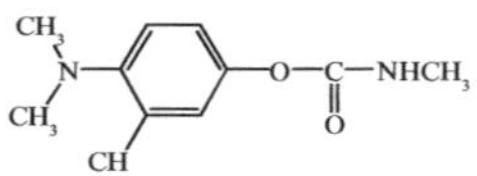	Foliar residues of conifer trees 90 g ai/ha, 0–150 d Acetonitrile wash & homogenates, GC	7 EC formulations with different diluents	Dislotogeable, DT_{50}=9–14 d Penetrated, DT_{50}=12–26 d	Sundaram (1986)

69	Deltamethrin 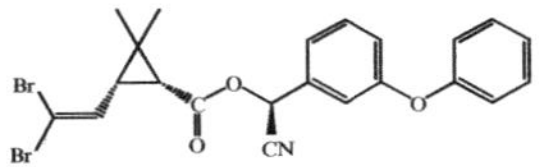	Foliar residues of staked cucumber 6 g ai/ha, 0–14 d EtOAc/Na dioctylsulfosinate extracts, GC	Decis® 25 EC Decis Ultra® 100 EC Decis® 200 CS	Dislodgeable residues, CS >> EC DT_{50} = 2.8 d (25EC), 3.8d (100EC), 32 d (200CS)	Franco et al. (2005)
70	Mepronil 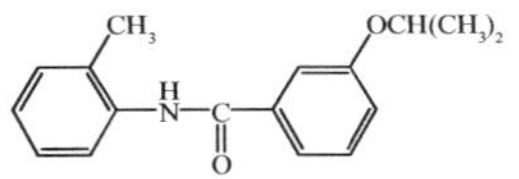	Residues in/on rice plants 150–180 g ai/10a, 0–24 d CH_3OH homogenates, CH_2Cl_2 extraction, GC	3% Dust 75% WP	Initial deposits, WP > D Insignificant differences in dissipation between D & WP	Asano et al. (1984)
17	Vinclozolin 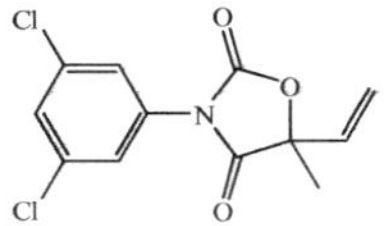	Foliar residues on peas, green house 0.19 mg ai/leaf, 0–45 d Water rinse, HPLC & acetone homogenates, GC	Acetone solution Ronilan® 50 WP	Surface rinse, WP >> ai DT_{50} (total) =33.1 d (WP), 13.4 d (ai) with no metabolites	Szeto et al. (1989)
49	Propiconazole 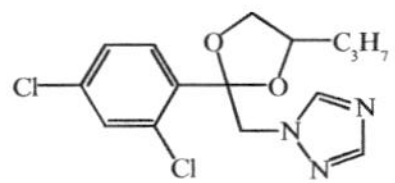	Residues on wheat & field beans 125 g ai/ha, 0–120 hr Ethyl acetate extraction, GC	Tilt® 25% EC + Agral® (1 mL/L) + Toil® (7.5 mL/L) + Bond® (1.4 mL/L)	Foliar dissipation, EC+Bond ≥ EC ≈ EC+Agral ≈ EC+Tol	Holloway and Western (2003)
71	Propargite 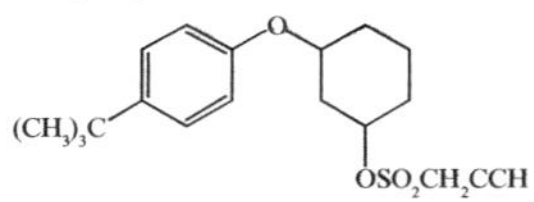	Foliar residues on apple trees 125 & 825 g ai/ha, 0–20 d Toluene extracts, GC	30 WP & 6 EC	Similar initial deposits. DT_{50} = 4.8–5.8 d (EC) 4.5–5.7 D (WP)	Marshall and Pree (1993)

(continued)

Table 11 (continued)

No.	Pesticide/structure	Residue trial/Rate, period/analysis	Formulation/additives	Effects	Reference
Soil residues					
65	Chlorpyrifos 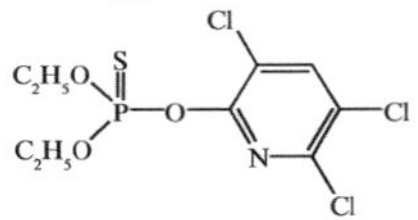	Citrus grove field 680–730 g ai/ha, 0–130 d Acetone extracts, GC	Clorpiran® 40 EC Dursban® 75 WG	DT_{50} = 7.6 d (EC), 6.3 d (WG)	Montemurro et al. (2002)
72	Temephos 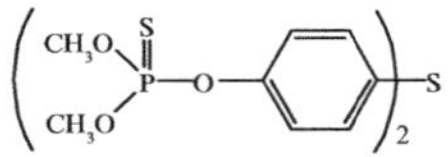	Agricultural farm (sandy loam) 1 kg ai/ha, 0–55 d CH_2Cl_2 extraction, colorimetry	50 EC & 1% G	DT_{50} = 18.1 d (EC) 20.4 d (G)	Verma et al. (2004)
73	Phenmedipham 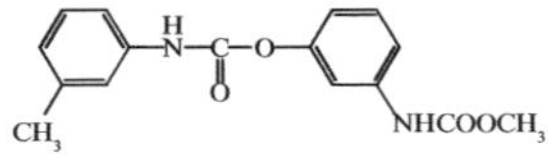	Sugar beet field 320 g ai/ha, 136 d HPLC	Commercial EC + Olbras®[d] + Atpolan®	Degradation rate, Commercial > Atpolan > Olbras	Kucharski et al. (2002)
74	Butachlor 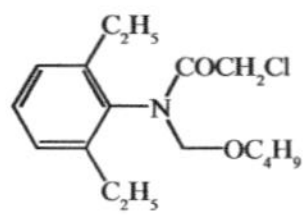	Rice field, pre-emergence 1–3 kg ai/ha, 0–115 d Acetonitrile extraction, GC	Machete® 50 EC Mon 12345 Machete® 50 EW	DT_{50} = 2.9–3.6 d 2.7–3.4 d 2.3–3.6 d	Kulshrestha (1987)
15	Trifluralin 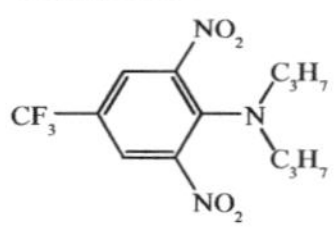	Agricultural field (loamy sand) 960 g ag/ha + 1% adjuvant, 57 d MeOH extraction, GC	Commercial EC + 5 adjuvants	Slower dissipation and more residues with adjuvants	Swarcewicz et al. (1998)

75	Fipronil 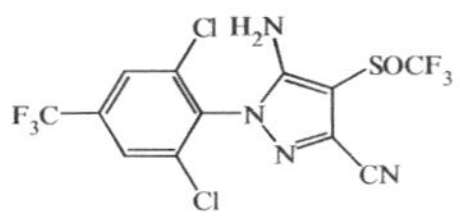	Rice field 28–560 g ai/ha, 0–400 hr CH_3OH & Solid-phase extraction GC	Icon® 80 WDG Icon® 1.67 SC Icon® 1.5 G	DT_{50} (water) = 0.4–5.2 d G > WDG > SC DT_{50} (soil) = 1.9–22.2 d WDG > G > SC	Ngim and Crosby (2001)
76	Oxyfluorfen 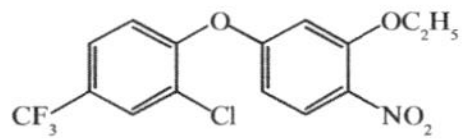	3 olive grove fields 480 g ai/ha, 0–160 d CH_3OH extraction, HPLC	Goal® 2XL (24 EC) Goal® G4F (48 SC)	DT_{50} = 11–21 d (EC) 16–23 d (SC)	Martínez et al. (2004)
13	Atrazine 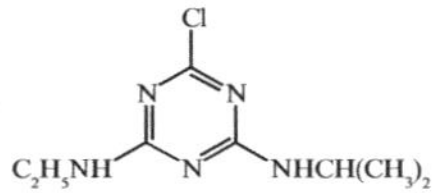	Corn field (clay loam) 1.68 kg ai/ha, 0–4 mon CH_3OH extraction, GC	Flowable (FL) WP	DT_{50} = 65 d (FL) 69 d (WP)	Khan et al. (1981)

[a] Mixture of amine salts of organic acids, aromatic acids and aromatic & aliphatic petroleum distillates.
[b] Agral, polyoxyethylene nonylphenols; Toil, methyl esters of rapeseed fatty acids; Bond, styrene-butadiene copolymers.
[c] Ortho HDD, alkylphenoxy polyethylene glycol.
[d] Olbras, postrefined fatty acids emulsifier; Atpolan, paraffin oil emulsifier; NA, not available.

The effect of formulation on soil dissipation has been reported for some pesticides. Much faster dissipation of azinphos-methyl (39) in the EC formulation than the granule (G) was observed in the field (Schulz et al. 1970). Since a similar trend was reported for the EC and G formulations of chlorpyrifos (65) in muck soil but with an insignificant difference in sandy soil (Chapman and Chapman 1986), the observed differences between formulations are likely at least in part to originate from different methods of application where less a.i. is desorbed from the granule. Ngim and Crosby (2001) clearly showed through field dissipation of fipronil (75) that less desorption of (75) from the G formulation than the soluble concentrate is the dominant reason for its longer persistence in soil. When the wettable granule formulation of (65) was used in field trials on oranges, a similar dissipation rate to the EC was reported (Montemurro et al. 2002). Insignificant differences in field dissipation rate among EC and other formulations were observed for atrazine (13) (Kahn et al. 1981), temephos (72) (Verma et al. 2004), butachlor (74) (Kulshrestha 1987), and oxyfluorfen (76) (Martínez et al. 2004). Dissipation of phenmedipham (73) applied to a sugar beet field was slightly reduced when refined fatty acid and paraffinic oil were used as adjuvants, to a larger extent in the former (Kucharski et al. 2002). The slower dissipation of EC formulations of several pesticides was also reported by addition of adjuvants (Kucharski 2003). Reduced volatilization by addition of adjuvants could account for the slower dissipation of trifluralin (15) as the EC (Swarcewicz et al. 1998).

IX Effects on Bioconcentration

Effect of adjuvants on bioconcentration of pesticide to the edible part of terrestrial plants is unlikely because of the insignificant increase of translocation from the applied sites, as discussed in Sections VII and VIII. White et al. (2007) reported the slightly increased bioaccumulation of DDE (1,1-dichloro-2,2-bis (*p*-chlorophenyl) ethylene) in the root and stem of a zucchini species but not in the fruit. They also reported the enhanced bioaccumulation of DDE to some earthworm species in the presence of Triton X-100, Tween 80, and rhamnolipids. In the case of fish, pesticides in water are considered to be taken up directly from water by gills or indirectly from food by the gastrointestinal tract. Because either more solubilization or less adsorption to solid matter is reported for pesticides in the presence of surfactant at concentration greater than cmc, the effect of adjuvants including surfactant on bioconcentration is anticipated. Aqueous humic acids having a solubilizing ability similar to surfactant are considered to reduce the bioavailability of a chemical, which would result in less apparent toxicity to fish (EPA 1996; Schrap 1991). The OECD testing guideline on fish bioconcentration does not recommend the use of solvents or dispersants, to avoid their toxicity and unexpected effects on bioconcentration, but some of them such as dimethylformamide, Tween 80, and HCO-40 are permitted at less than 100 mg/L if unavoidable because of the low water solubility of a test chemical (OECD 1996). Because most regulations only require the fish bioconcentration for a hydrophobic chemical with log *P* greater than 3 and most studies have been conducted in the presence of some solvents or dispersants, information on the effect of adjuvants on the bioconcentration of a chemical are very limited. Kaka and Hayton (1978) studied the effect of 0.01% polysorbate 80 on the bioaccumulation of 4-aminoantipyrine and ethanol in goldfish at 10°–35°C and reported increase of uptake rate, especially for the former compound, presumably from enhanced diffusion through the gill. In contrast, any significant effect of nonionic surfactant at the concentrations less than cmc was not observed in a recent study.

The bioconcentrations of seven hydrophobic chemicals (log *P* = 4.2–6.9) including DDT (1) to carp have been investigated in the presence and absence of castor oil HCO surfactants having the polyethyleneglycol substructure in accordance with OECD 305, and almost similar bioconcentration factors were obtained irrespective of the nonionic surfactants (Yakata et al. 2006). Alternatively, Park et al. (2002) demonstrated the significantly reduced bioconcentration of three PAHs to freshwater mussels in the presence of 0.02% Tween 80 above cmc. They assumed the partition of PAH to the micelles and obtained micelle partition coefficients of $5–6 \times 10^3$ (L/kg) by analyzing the time-dependent uptake profiles in the presence of the surfactant.

Effects of the dispersant Corexit 9527 including ethoxylated sorbitans, anionic surfactant, and hydrocarbon solvent on the bioconcentration of PAH in a few aquatic species have been investigated extensively as well as its trophic transfer. Wolfe et al. (1998a,b) reported not only increased bioaccumulation of naphthalene to marine algae at 20°C but also increased trophic transfer to the rotifer by the dispersant at 20 ppm. The dispersant-induced leakiness and lysis of algal cells resulted in the increased depuration of naphthalene as a result of the increased cell permeability. Similar effects of this dispersant were also reported for phenanthrene (Wolfe et al. 1999). By using these PAHs in the presence of Corexit 9527, their uptake and trophic transfer in the model aquatic system including algae (primary producer), rotifer (primary consumer), and laval topsmelt (higher consumer) were observed (Wolfe et al. 2001; Mielbrecht et al. 2005). The dispersant increased the uptake and trophic transfer of less hydrophobic naphthalene, but the bioconcentration of phenanthrene in topsmelt decreased, probably from more solubilization of the PAH in the micelles. Although available information on the effect of adjuvants and surfactants is limited, not only their concentration above or below cmc but also the hydrophobicity of a chemical are the controlling factors for bioconcentration in aquatic species.

Summary

The potential effects of adjuvants, including surfactants used in pesticide formulation, have been extensively studied for many small organic chemicals, but similar investigation on pesticides is limited in most cases. Solubilizing effects leading to the apparently increased water solubility of a pesticide are commonly known through the preparation of formulations, but fundamental profiles, especially for a specific monodisperse surfactant, are not fully studied. Reduced volatilization of a pesticide from the formulation can be explained by analogy of a very simple organic chemical, but the actual mechanism for the pesticide is still obscure. In contrast, from the point of view of avoiding groundwater contamination with a pesticide, adsorption/desorption profiles in the presence of surfactants and adjuvants have been examined extensively as well as pesticide mobility in the soil column. The basic mechanism in micelle-catalyzed hydrolysis is well known, and theoretical approaches including the PPIE model have succeeded in explaining the observed effects of surfactants, but its application to pesticides is also limited. Photolysis, especially in an aqueous phase, is in the same situation. The dilution effect in the real environment would show these effects on hydrolysis and photolysis to be much less than expected from the laboratory basic studies, but more information is necessary to examine the practical extent of the effects in an early stage of applying a pesticide formulation to crops and soil. Many adjuvants, including surfactants, are biodegradable in the soil environment, and thus their effects on the biodegradation of a pesticide in soil and sediment may be limited, as demonstrated by field trials. Not only from the theoretical but also the practical aspect, the foliar uptake of pesticide in the presence of adjuvants has been investigated extensively and some prediction on the ease of foliar uptake can be realized in relation to the formulation technology. However, effect on root uptake of pesticides is to be further investigated. In an aqueous environment more or less contaminated by various chemicals such as detergents and their degradates, it is necessary to investigate the effect of adjuvants on uptake, bioconcentration, and trophic transfer of pesticides for better understanding of pesticide contamination of aquatic species in the aquatic environment.

References

Abu-Hamdiyyah A, Rahman IA (1985) Strengthening of hydrophobic binding and the increase in the degree of micellar ionization by amphiphiles and the micelle-water distribution coefficient as a function of the surfactant chain length in sodium alkyl sulfates. J Phys Chem 89:2377–2384.

Abu-Hamdiyyah A, Rahman IA (1987) Distribution coefficients of nonpolar additives and factors determining the solubilization tendency as a function of surfactant chain length in aqueous solution of sodium alkyl sulfates. J Phys Chem 91:1530–1535.

Abu-Zreig M, Rudra RP, Dickinson WT (2000) Influence of surfactants on leaching of atrazine through soil columns. Toxicol Environ Chem 75:1–16.

Ahmad R, Kookana RS, Alston AM (2004) Surfactant-enhanced release of carbaryl and ethion from two long-term contaminated soils. J Environ Sci Health 39B:565–576.

Al-Awadi N, Williams A (1990) Effective charge development in ester hydrolysis by cationic micelles. J Org Chem 55:2001–2004.

Allen CCR, Boyd DR, Hempenstall F, Larkin MJ, Sharm ND (1999) Contrasting effects of a nonionic surfactant on the biotransformation of polycyclic aromatic hydrocarbons to *cis*-dihydrodiols by soil bacteria. Appl Environ Microbiol 65:1335–1339.

Almgren M, Grieser F, Thomas JK (1980) Photochemical and photophysical studies of organized assemblies. Interaction of oils, long-chain alcohols, and surfactants forming microemulsions. J Am Chem Soc 102:3188–3193.
Amonette J, O'Connor GA (1990) Nonionic surfactant effects on adsorption and degradation of 2,4-D. Soil Sci Soc Am J 44:540–544.
Anderson MA (1992) Influence of surfactants on vapor–liquid partitioning. Environ Sci Technol 26:2186–2191.
Anderson NH, Girling J (1983) The uptake of surfactants into wheat. Pestic Sci 14:399–404.
Anvir D, Johnston LJ, DeMayo P, Wong SK (1981) Surface photochemistry: radical pair combination on a silica gel surface and in micelles. J Chem Soc Chem Commun 958–959.
Arias M, García-Río L, Mejuto JC, Rodríguez-Dafonate, Simal-Gándara J (2005) Influence of micelles on the basic degradation of carbofuran. J Agric Food Chem 53:7172–7178.
Aronstein BN, Calvillo YM, Alexander M (1991) Effect of surfactants at low concentrations on the desorption and biodegradation of sorbed aromatic compounds in soil. Environ Sci Technol 25:1728–1731.
Asano Y, Oishi T, Abe H, Anma K, Ishikawa K, Ishihara E (1984) Persistence of mepronil in rice plants treated with two formulations of the fungicide. J Pestic Sci 9:643–649.
Atwood ST, Sheets TJ, Sutton TB, Leidy RB (1987) Stability of selected pesticide formulations and combinations in aqueous media. J Agric Food Chem 35:169–172.
Awasthi N, Kumar A, Makkar R, Cameotra SS (1999) Biodegradation of soil-applied endosulfan in the presence of a biosurfactant. J Environ Sci Health B34:793–803.
Baker EA, Hunt GM, Stevens PJG (1983) Studies of plant cuticle and spray droplet interactions: a fresh approach. Pestic Sci 14:645–658.
Balakrishnan VK, Han X, VanLoon GW, Dust JM, Toullec J, Buncel E (2004) Acceleration of nucleophilic attack on an organophosphorothioate neurotoxin, fenitrothion, by reactive counterion cationic micelles. Regioselectivity as a probe of substrate orientation within the micelle. Langmuir 20:6586–6593.
Balakrishnan VK, Buncel E, VanLoon GW (2005) Micellar catalyzed degradation of fenitrothion, an organophosphorus pesticide, in solutions and soils. Environ Sci Technol 39:5824–5830.
Baloch R (2000) Experimental approaches for plant metabolism studies. In: Roberts T (ed) Metabolism of Agrochemicals in Plants. Wiley, New York, pp 5–42.
Barron MG, Little EE, Calfee R, Diamonds S (2000) Quantifying solar spectral irradiance in aquatic habitats for the assessment of photoenhanced toxicity. Environ Toxicol Chem 19:920–925.
Bartlett JR, Cooney RP (1986) Raman spectroscopic studies of poly(oxyethylene) chains conformations in non-ionic surfactants. J Chem Soc Faraday Trans I 82:597–605.
Bauer H, Schönherr J (1992) Determination of mobilities of organic compounds in plant cuticles and correlation with molar volumes. Pestic Sci 35:1–11.
Baur P (1998) Mechanistic aspects of foliar penetration of agrochemicals and the effect of adjuvants. Recent Res Dev Agric Food Chem 2:809–837.
Bayer DE (1967) Effects of surfactants on leaching of substituted urea herbicides in soil. Weeds 15:249–252.
Beestman GB, Deming JM (1976) Triallate mobility in soil. Weed Sci 24:541–544.
Behzadipour M, Luge MK, Lüthje S (2001) Changes in plasma membrane fluidity of corn (*Zea mays* L.) roots after Brij 58 treatment. Protoplasma 217:65–69.
Beigel C, Barriuso E (2000) Influence of formulation on triticonazole solubilization and sorption in a soil–water system. Pestic Manag Sci 56:271–276.
Beigel C, Charnay MP, Barriuso E (1999) Degradation of formulated and unformulated triticonazole fungicide in soil: effect of application rate. Soil Biol Biochem 31:525–534.
Beme MTA, Fullington JG, Noel R, Cordes EH (1965) Secondary valence force catalysis. II. Kinetics of the hydrolysis of orthoesters and the hydrolysis and aminolysis of carboxylic esters in the presence of micelle-forming detergents. J Am Chem Soc 87:266–270.
Bentson KP (1990) Fate of xenobiotics in foliar pesticide deposits. Rev Environ Contam Toxicol 114:125–161.

Bernardez LA, Ghoshal S (2004) Selective solubilization of polycyclic aromatic hydrocarbons from multicomponent nonaqueous-phase liquids into nonionic surfactant micelles. Environ Sci Technol 38:5878–5887.
Bertolotti SG, García NA, Gsponer HE (1989a) Micellar binding of polychlorophenols. J Colloid Interface Sci 129:406–413.
Bertolotti SG, Gsponer HE, García NA (1989b) Surfactant effect on the sensitized photooxidation of polychlorophenolic pesticides. Toxicol Environ Chem 22:229–237.
Bokern M, Harms HH (1997) Toxicity and metabolism of 4-*n*-nonylphenol in cell suspension cultures of different plant species. Environ Sci Technol 31:1849–1859.
Bokern M, Nimtz M, Harns HH (1996) Metabolites of 4-*n*-nonylphenol in wheat cell suspension cultures. J Agric Food Chem 44:1123–1127.
Bokern M, Raid P, Harms HH (1998) Toxicity, uptake and metabolism of 4-*n*-nonylphenol in root cultures and intact plants under septic and aseptic conditions. Environ Pollut Res 5:21–27.
Bowman BT (1993) Effect of formulation upon movement and dissipation of ^{14}C-metolachlor and atrazine in field lysimeters. Can J Soil Sci 73:309–316.
Boyd SA, Lee JF, Mortland MM (1988) Attenuating organic contaminant mobility by soil modification. Nature (Lond) 333:345–347.
Briggs GG, Bromilow RH, Evans AA (1982) Relationship between lipophilicity and root uptake and translocation of non-ionized chemicals by barley. Pestic Sci 13:495–504.
Briggs GG, Bromilow RH, Evans AA, Williams M (1983) Relationship between lipophilicity and the distribution of non-ionized chemicals in barley shoots following uptake by root. Pestic Sci 14:492–500.
Bromilow RH, Evans AA, Nicholls PH (2003) The influence of lipophilicity and formulation on the distribution of pesticides in laboratory-scale sediment/water systems. Pestic Manag Sci 59:238–244.
Brown DG (2007) Relationship between micellar and hemi-micellar processes and the bioavailability of surfactant-solubilized hydrophobic organic compounds. Environ Sci Technol 41:1194–1199.
Brown DG, Jaffé PR (2006) Effects of nonionic surfactants on the cell surface hydrophobicity and apparent Hamaker constant of a *Sphingomonas* sp. Environ Sci Technol 40:195–201.
Broxton TJ (1981) Micellar catalysis of organic reactions. VII. The effect of the micellar counter ion in nucleophilic reactions of hydroxide and nitrite ions. Aust J Chem 34:2313–2319.
Broxton TJ (1983) Basic hydrolysis of some alkyl and phenyl *N*-aryl-*N*-methylcarbamates. Aust J Chem 36:2203–2209.
Broxton TJ (1984) Micellar catalysis of organic reactions. XII. Basic hydrolysis of some alkyl and aryl *N*-(4-nitrophenyl)carbamates. Aust J Chem 37:47–54.
Broxton TJ, Chung RPT (1986) Micellar catalysis of organic reactions. 19. Basic hydrolysis of carbamates in the presence of hydroxyl-functionalized micelles. J Org Chem 51:3112–3115.
Broxton TJ, Duddy NW (1979) Substituent effects on the micelle-catalyzed and uncatalyzed basic hydrolysis of a series of substituted *N*-methyl-*p*-toluanilides. Aust J Chem 32:1717–1726.
Broxton TJ, Duddy NW (1980) Micellar catalysis of the basic hydrolysis of anilides. III α-Substituted *N*-methyl-*N*-*p*-nitrophenylacetamides. Aust J Chem 33:1771–1781.
Broxton TJ, Sango DB (1983) Micellar catalysis of organic reactions. X. Further evidence for the partial failure of the pseudophase kinetic model of micellar catalysis for reactions of hydroxide ions. Aust J Chem 36:711–717.
Broxton TJ, Deady LW, Duddy NW (1978) Micellar catalysis of anilide hydrolysis. Aust J Chem 31:1525–1532.
Broxton TJ, Fernando DR, Rowe JE (1981) Micellar catalysis of the basic hydrolysis of amides. 4. Substituted *N,N*-diphenylbenzamides. J Org Chem 46:3522–3525.
Broxton TJ, Christie JR, Sango X (1987) Micellar catalysis of organic reactions. 20. Kinetic studies of the hydrolysis of aspirin derivatives in micelles. J Org Chem 52:4814–4817.
Broxton TJ, Christie JR, Mannas SM (1988) Micellar catalysis of organic reactions. XXI. A comparison of the catalytic activity of micelles of cetyltrimethylammonium bromide and sulfate on ester, amide and carbamate hydrolyses. Aust J Chem 41:325–330.

Bukovac MJ, Cooper JA, Whitmoyer RE, Brazee RD (2003) Pesticide delivery: multiple role of adjuvants in foliar application of systemic compounds. ASTM Special Technical Publication 1430. American Society for Testing and Materials, Philadelphia, pp 91–107.
Bunton CA, Minch MJ (1974) Micellar effects on the ionization of carboxylic acids and interactions between quaternary ammonium ions and aromatic compounds. J Phys Chem 78:1490–1498.
Bunton CA, Robinson L (1969) Micellar effects upon the reaction of *p*-nitrophenyl diphenyl phosphate with hydroxide and fluoride ion. J Org Chem 34:773–780.
Bunton CA, Savelli G (1986) Organic reactivity in aqueous micelles and similar assemblies. Adv Phys Org Chem 22:213–309.
Bunton CA, Fendler EJ, Sepulveda L, Yang KU (1968) Micellar-catalyzed hydrolysis of nitrophenyl phosphates. J Am Chem Soc 90:5512–5518.
Bunton CA, Robinson L, Sepulveda L (1969) Micellar effects upon phosphorylation and phosphate ester hydrolysis. J Am Chem Soc 91:4813–4819.
Bunton CA, Romsted LS, Savelli G (1979) Tests of the pseudophase model of micellar catalysis: Its potential failure. J Am Chem Soc 101:1253–1259.
Bunton CA, Romsted LS, Sepulveda L (1980) A quantitative treatment of micellar effects upon deprotonation equilibria. J Phys Chem 84:2611–2618.
Bunton CA, Nome F, Quina FH, Romsted LS (1991) Ion binding and reactivity at charged aqueous interfaces. Acc Chem Res 24:357–364.
Burghardt M, Schreiber L, Riederer M (1998) Enhancement of the diffusion of active ingredients in barley leaf cuticular wax by monodisperse alcohol ethoxylates. J Agric Food Chem 46:1593–1602.
Burghardt M, Friedmann A, Schreiber L, Riederer M (2006) Modelling the effects of alcohol ethoxylates on diffusion of pesticides in the cuticular wax of *Chenopodium album* leaves. Pestic Manag Sci 62:137–147.
Bury SJ, Miller CA (1993) Effect of micellar solubilization on biodegradation rates of hydrocarbons. Environ Sci Technol 27:104–110.
Caboni P, Cabras M, Angioni A, Russo M, Cabras P (2002) Persistence of azadirachtin residues on olives after field treatment. J Agric Food Chem 50:3491–3494.
Cano ML, Dorn PB (1996) Sorption of an alcohol ethoxylate surfactant to natural sediments. Environ Toxicol Chem 15:684–690.
Cao Y, Chen J, Wang Y, Liang J, Chen L, Lu Y (2005) HPLC/UV analysis of chlorfenapyr residues in cabbage and soil to study the dynamics of different formulations. Sci Total Environ 350:38–46.
Celis R, Hermosín MC, Carrizosa MJ, Cornejo J (2002) Inorganic and organic clays as carriers for controlled release of the herbicide hexazinone. J Agric Food Chem 50:2324–2330.
Chang-Yen I, Nickless G, Pickard JA (1983) Effect of pesticide formulation on the distribution of fenitrothion on apple foliage. J Agric Food Chem 31:1137–1139.
Chao WL, Lin CM (2006) Degradation of dibutyl-phthalate by soil bacteria. Chemosphere 63:1377–1383.
Chaplain V, Barriuso E, Dur JC, Vergnet C (2001) Influence of the formulation on the sorption and the mobility of diuron in soil. Bull Environ Contam Toxicol 66:664–670.
Chapman RA, Chapman PC (1986) Persistence of granular and EC formulations of chlorpyrifos in a mineral and an organic soil incubated in open and closed containers. J Environ Sci Health B21:447–456.
Chappell MA, Laird DA, Thompson ML, Evangelou VP (2005) Cosorption of atrazine and a lauryl polyoxyethylene oxide nonionic surfactant on smectite. J Agric Food Chem 53:10127–10133.
Charnay MP, Tarabelli L, Beigel C, Barriuso E (2000) Modifications of soil microbial activity and triticonazole biodegradation by pesticide formulation additives. J Environ Qual 29:1618–1624.
Chattopadhyay K, Das TK, Majumdar A, Mazumdar S (2002) NMR studies on interaction of lauryl maltoside with cytochrome C oxidase: a model for surfactant interactions with the membrane protein. J Inorg Biochem 91:116–124.

Chen JL, Horne PA, Jackson WR, Lichti G, Park D (1994) Volatility control for foliage-applied chlorpyrifos by using controlled release formulations. J Controlled Release 29:83–95.
Chen L, Zhou HY, Deng QY (2007) Photolysis of nonylphenol ethoxylates: the determination of the degradation kinetics and the intermediate products. Chemosphere 68:354–359.
Cheng KY, Wong JWC (2006a) Effect of synthetic surfactants on the solubilization and distribution of PAHs in water-soil systems. Environ Technol 27:835–844.
Cheng KY, Wong JWC (2006b) Combined effect of nonionic surfactant Tween 80 and DOM on the behaviors of PAHs in soil-water system. Chemosphere 62:1907–1916.
Chiou CT, Kile DE, Rutherford DW (1991) The neutral oil in commercial lonear alkylbenzenesulfonate and its effect on organic solute solubility in water. Environ Sci Technol 25:660–665.
Cho HH, Park JW, Liu CCK (2002) Effect of molecular structure on the solubility enhancement of hydrophobic organic compounds by environmental amphiphiles. Environ Toxicol Chem 21:999–1003.
Chowdary KPR, Rao SKS (2001) Effect of surfactants on the solubility and dissolution rate of itraconazole. East Pharm 44:121–123.
Choy WK, Chu W (2001) The modeling of trichloroethane photodegradation in Brij 35 surfactant by two-stage reaction. Chemosphere 44:211–215.
Chu W (1999) Photodechlorination mechanism of DDT in a UV/surfactant syetm. Environ Sci Technol 33:421–425.
Chu W, Jafvert CT (1994) Photodechlorination of polychlorobenzene congeners in surfactant micelle solutions. Environ Sci Technol 28:2415–2422.
Chu W, Jafvert CT, Diehl CA (1998) Phototransformations of polychlorobiphenyls in Brij 58 micellar solutions. Environ Sci Technol 32:1989–1993.
Copping L (2000) Adjuvants and additives in crop protection. AGROW Report DS 183. PJB Publications, Surrey.
Cordes EH, Gitler C (1972) Reaction kinetics in the presence of micelle-forming surfactants. Prog Bioorg Chem 2:1–53.
Coret J, Chamel A (1995) Effects and possible mode of action of some nonionic surfactants on the diffusion of [^{14}C] chlorotoluron across isolated plant cuticles. Pestic Sci 43:163–180.
Cotterill EG (1988) The effect of formulation as suspendable concentrate or wettable powder on degradation, adsorption and concentration in soil solution of chlorotoluron. Chemosphere 17:1041–1047.
Crocker FH, Guerin WF, Boyd SA (1995) Bioavailability of naphthalene sorbed to cationic surfactant-modified smectite clay. Environ Sci Technol 29:2953–2958.
Cserháti T (1995) Alkyl ethoxylated and alkylphenol ethoxylated nonionic surfactants: interaction with bioactive compounds and biological effects. Environ Health Perspect 103:358–364.
Da Silva JP, Da Silva AM, Khmelinskii IV (2001) Dissipation of triadimefon on the solid/gas interface. Chemosphere 45:875–880.
Dailey OD Jr (2004) Volatilization of alachlor from polymeric formulations. J Agric Food Chem 52:6742–6746.
Danikhel RK, Purnanand (1990) Micellar catalysis of hydroperoxide anion assisted hydrolysis of *O-p*-cyanophenyl *O*-ethyl phenyl phosphonothionate. Indian J Chem 29A:856–860.
Davis L, Bonner F, Hensley SD (1969) Residues on sugarcane tissues 24 hours after application of different formulations of azinphos methyl. J Econ Entomol 62:505–506.
de Castro B, Domingues V, Gameiro P, Lima JLFC, Oliveira A, Reis A (1999) Acid-base properties and solubility of pindolol, diazepam and chlordiazepoxide in SDS micelles. Int J Pharm 187:67–75.
de Ruiter H, Mol HGJ, de Vliger JJ, van de Zande JC (2003) Proceedings, BCPC International Congress, Crop Science & Technology, vol 1, Glasgow, pp 513–518.
Diallo MS, Abriola LM, Weber Jr WJ (1994) Solubilization of nonaqueous phase liquid hydrocarbons in micellar solutions of dodecyl alcohol ethoxylates. Environ Sci Technol 18:1829–1837.
Di Cesare D, Smith JA (1994) Surfactant effects on desorption rate of nonionic organic compounds from soil to water. Rev Environ Contam Toxicol 134:1–29.

Di Toro DM, Dodge LJ, Hand VC (1990) A model for anionic surfactant sorptions. Environ Sci Technol 24:1013–1020.
Doucette WJ, Wheeler BR, Chard JK, Bugbee B, Naylor CG, Carbone JP, Sims RC (2005) Uptake of nonylphenol and nonylphenol ethoxylates by crested wheatgrass. Environ Toxicol Chem 24:2965–2972.
Drummond CJ, Grieser F, Healy TW (1989) Acid–base equilibria in aqueous micellar solutions. Part 1. Simple weak acids and bases. J Chem Soc Faraday Trans I 85:521–535.
Edwards DA, Luthy RG, Liu Z (1991) Solubilization of polycyclic aromatic hydrocarbons in micellar nonionic surfactant solutions. Environ Sci Technol 25:127–133.
Edwards DA, Liu Z, Luthy RG (1992) Interactions between nonionic surfactant monomers, hydrophobic organic compounds and soil. Water Sci Technol 26:147–158.
Edwards DA, Adeel Z, Luthy RG (1994) Distribution of nonionic surfactant and phenanthrene in a sediment/aqueous system. Environ Sci Technol 28:1550–1560.
Eichhorn P (2003) Occurrence of surfactants in surface waters and freshwater sediments. II. Linear alkylbenzene sulfonates and their carboxylated degradation products. In: Knepper TP, Barceló D, de Voogt P (eds) Comprehensive Analytical Chemistry, vol 40. Elsevier, New York, pp 695–718.
Ekler Z (1988) Behavior of thiocarbamate herbicides in soils: adsorption and volatilization. Pestic Sci 22:145–157.
El-Nahhal Y, Nir S, Polubesova T, Margulies L, Rubin B (1998) Leaching, phytotoxicity and weed control of new formulation of alachlor. J Agric Food Chem 46:3305–3313.
Elsgaard L, Petersen SO, Debosz K (2001) Effects and risk assessment of linear alkylbenzene sulfonates in agricultural soil. 1. Short-term effects on soil microbiology. Environ Toxicol Chem 20:1656–1663.
EPA (1996) OPPTS Harmonized Test Guidelines. Ecological Effects Guidelines. OPPTS 850.1085. Fish acute toxicity mitigated by humic acid (public draft). Office of Prevention, Pesticides and Toxic Substances, U.S. Environmental Protection Agency, Washington, DC.
Eriksson JC, Gillberg G (1966) NMR-studies of the solubilization of aromatic compounds in cetyltrimethylammonium bromide solution II. Acta Chem Scand 20:2019–2027.
Esumi K, Takeda Y, Koide Y (1998) Competitive adsorption of cationic surfactant and pesticide on laponite. Colloids Surfaces A Physicochem Eng Aspects 135:59–62.
Falb LN, Bridges DC, Smith AE Jr (1990) Effects of pH and adjuvants on clethodim photodegradation. J Agric Food Chem 38:875–878.
Fendler JH, Bogan G (1974) Micellar effects on uracil photodimerization. Photochem Photobiol 20:323–328.
Fendler JH, Fendler EJ (1975) Catalysis in Micellar and Macromolecular Systems. Academic Press, New York.
Fendler JH, Fendler EJ, Infante GA, Shih PS, Patterson LK (1975) Absorption and proton magnetic resonance spectroscopic investigation of the environment of acetophenone and benzophenone in aqueous micellar solutions. J Am Chem Soc 97:89–95.
Field RJ, Dastgheib F (1996) Enhancing uptake and translocation of systemic active ingredients. In: Foy CL, Pritchard DW (eds) Pesticide Formulation and Adjuvant Technology. Formulation Forum '94, Washington, DC. CRC Press, Boca Raton, pp 241–295.
Forster WA, Zabkiewicz JA, Riederer M (2004) Mechanisms of cuticular uptake of xenobiotics into living plants: 1. Influence of xenobiotic dose on the uptake of three model compounds applied in the absence and presence of surfactants into *Chenopodium album*, *Hedera helix* and *Strephanotis floribunda* leaves. Pestic Manag Sci 60:1105–1113.
Forster WA, Zabkiewicz JA, Liu Z (2006a) Cuticular uptake of xenobiotics into living plants. Part 2: Influence of the xenobiotic dose on the uptake of bentazone, epoxiconazole and pyraclostrobin, applied in the presence of various surfactants, into *Chenopodium album*, *Sinapis alba* and *Triticum aestivum* leaves. Pestic Manag Sci 62:664–672.
Forster WA, Zabkiewicz JA, Riederer M (2006b) Mechanisms of cuticular uptake of xenobiotics into living plants: evaluation of a logistic-kinetic penetration model. J Agric Food Chem 54:3025–3032.

Foy CL (1992) Influence of certain surfactants on the mobility of selected herbicides in soil. In: Foy CL (ed) Adjuvants for Agrichemicals. CRC Press, Boca Raton, pp 349–366.
Franco AA, Trevizan LRP, de Baptista GC, Cassamassimo RE (2005) Leaf dislodgeable residues of deltamethrin applied in different types of formulations in staked cucumbers. Rev Brasil Toxicol 18:5–10.
Gao B, Wang X, Zhao J, Sheng G (2001) Sorption and cosorption of organic contaminant on surfactant-modified soils. Chemosphere 43:1095–1102.
Gao Y, Ling W, Wong MH (2006) Plant-accelerated dissipation of phenanthrene and pyrene from water in the presence of a non-ionic surfactant. Chemosphere 63:1560–1567.
García-Ortega S, Holliman PJ, Jones DL (2006) Toxicology and fate of Pestanal and commercial propetamphos formulation in river and estuarine sediment. Sci Total Environ 366:826–836.
Garratt JA, Wilkins RM (2000) Assessment of volatilization rates of pesticides from formulations. In: Cornejo J, Jamet P (eds) Pesticide/Soil Interactions: Some Current Research Methods. INRA, Paris, pp 43–47.
Gavril D, Atta KR, Karaiskakis G (2006) Study of the evaporation of pollutant liquids under the influence of surfactants. Am Inst Chem Eng J 52:2381–2390.
Gerakis AM, Koupparis MA, Efstathiou CE (1993) Micellar-acid-base potentiometric titrations of weak acidic and/or insoluble dyes. J Pharm Biomed Anal 11:33–41.
Gong A, Ye C, Wang X, Lei Z, Liu J (2001) Dynamics and mechanism of ultraviolet photolysis of atrazine on soil surface. Pestic Manag Sci 57:380–385.
Gonsalves M, Probst S, Rezende MC, Nome F, Zucco D (1985) Failure of the pseudophase model in the acid-catalyzed hydrolysis of acetals and *p*-methoxybenzaldoxime esters in the presence of an anionic micelles. J Phys Chem 89:1127–1130.
Gonzalez V, Ayala JH, Afonso AM (1992) Degradation of carbaryl in natural waters: Enhanced hydrolysis rate in micellar solution. Bull Environ Contam Toxicol 48:171–178.
Grant RJ (2001) Biodegradation of synthetic pyrethroid insecticides in formulated compounds. In: Magar VS, von Fahnestock FM, Lecson A (eds) Ex Situ Biological Treatment Technologies. The 6th International In Situ and On-Site Bioremediation Symposium, San Diego. Battelle Press, Columbus, pp 27–34.
Green JM, Hale T (2005) Increasing and decreasing pH to enhance the biological activity of nicosulfuron. Weed Technol 19:468–475.
Green SA, Williams RJ, Stock D (1998) Influence of formulants and salt formation on volatilization and activity of pyrimethanil. Pestic Sci 54:313–314.
Gregoritch SJ, Thomas JK (1980) Photochemistry in microemulsions. Photophysical studies in oleate/hexanol/hexadecane, oil in water microemulsion. J Phys Chem 84:1491–1495.
Grieser F, Drummond CJ (1988) The physicochemical properties of self-assembled surfactant aggregates as determined by some molecular spectroscopic techniques. J Phys Chem 92:5580–5593.
Griffin WC (1949) Classification of surface-active agents by HLB. J Soc Cosmetic Chem 1:311–326.
Grimberg SJ, Nagel J, Aitken MD (1995) Kinetics of phenanthrene dissolution into water in the presence of nonionic surfactants. Environ Sci Technol 29:1480–1487.
Grimberg SJ, Stringfellow WT, Aitken MD (1996) Quantifying the biodegradation of phenanthrene by *Pseudomonas stutzeri* P16 in the presence of a nonionic surfactant. Appl Environ Microbiol 62:2387–2392.
Guerin WF, Jones GE (1988) Mineralization of phenanthrene by a *Mycobacterium* sp. Appl Environ Microbiol 54:937–944.
Guha S, Jaffé PR (1996a) Biodegradation kinetics of phenanthrene partitioned into the micellar phase of nonionic surfactants. Environ Sci Technol 30:605–611.
Guha S, Jaffé PR (1996b) Bioavailability of hydrophobic compounds partitioned into the micellar phase of nonionic surfactants. Environ Sci Technol 30:1382–1391.
Guha S, Jaffé PR, Peters CA (1998) Solubilization of PAH mixtures by a nonionic surfactant. Environ Sci Technol 32:930–935.
Günther FA (1969) Insecticide residues in California citrus fruits and products. Res Rev 28:1–119.

Günther P, Pestemer W (1992) Phytotoxicity of surfactants to higher plants. In: Report EUR 14236: Effects of organic contaminants in sewage sludge on soil fertility, plants and animals. Commission of the European Communities, Luxenbourg, pp 103–111.

Gupta S, Gajbhiye VT, Kalpana, Agnihotri NP (2002) Leaching behavior of imidacloprid formulations in soil. Bull Environ Contam Toxicol 68:502–508.

Haigh SD (1996) A review of the interaction of surfactants with organic contaminants in soil. Sci Total Environ 185:161–170.

Halle AT, Drncova D, Richard C (2000) Phototransformation of the herbicide sulcotrione on maize curricular wax. Environ Sci Technol 40:2989–2995.

Han X, Balakrishnam VK, van Loon GW, Buncel E (2006) Degradation of the pesticide fenitrothion as mediated by cationic surfactants and α-nucleophilic reagents. Langmuir 22:9009–9017.

Hand VC, Williams GK (1987) Structure-activity relationships for sorption of linear alkylbenzenesulfonates. Environ Sci Technol 21:370–373.

Hand VC, Rapaport RA, Wendt RH (1990) Adsorption of dodecyltrimethylammmonium chloride (C_{12}TMAC) to river sediment. Environ Toxicol Chem 9:467–471.

Hao J (2000) Studies on chemical reactions in microemulsion media (I). Hydrolysis kinetics studies in microemulsion system for CTAB/*n*-butanol/25% *n*-octane/water. J Dispers Sci Technol 21:19–30.

Harrison SK, Wax LM (1985) The effect of adjuvants and oil carriers on photodecomposition of 2,4-D, bentazone and haloxyfop. Weed Sci 34:81–87.

Harrison RB, Holmes DC, Roburn J, Tatton JOG (1967) The fate of some organochlorine pesticides on leaves. J Sci Food Agric 18:10–15.

Hautala RR (1978) Surfactant effects on pesticide photochemistry in water and soil. EPA Technical Report Data. EPA-600/3-78-060 (PB-285175). U.S. Environmental Protection Agency, Washington, DC, pp 1–83.

Hautala RR, Letsinger RL (1971) Effects of micelles on the efficiency of photoinduced substitution reactions and fluorescence quenching. J Org Chem 36:3762–3768.

Hazen JL (2000) Adjuvants: terminology, classification, and chemistry. Weed Technol 14:773–784.

Hazen JL, Krebs PJ (1992) Photodegradation and absorption of sethoxydim as adjuvant-influenced surface effects. In: Foy CL (ed) Adjuvants for Agrichemicals. CRC Press, Boca Raton, pp 195–203.

Helling CS (1971) Pesticide mobility in soils. II. Application of soil thin-layer chromatography. Soil Sci Soc Am Proc 35:737–743.

Heredia A, Bukovac MJ (1992) Interaction between 2-(1-naphthyl)acetic acid and micelles of nonionic surfactants in aqueous solution. J Agric Food Chem 40:2290–2293.

Hermosín MC, Celis R, Facenda G, Carrizosa MJ, Ortega-Calro JJ, Cornejo J (2006) Bioavailability of the herbicide 2,4-D formulated with organoclays. Soil Biol Biochem 38:2117–2124.

Hess FD, Foy CL (2000) Interaction of surfactants with plant cuticles. Weed Technol 14:807–813.

Hoffmann U, Gao Y, Pandy B, Klinge S, Warzecha KD, Krüger C, Roth HD, Demuth M (1993) Light-induced polyene cyclization via radical cations in micellar medium. J Am Chem Soc 115:10358–10359.

Holloway PG, Silcox D (1985) Behavior of three nonionic surfactants following foliar application. Proceedings, BCPC Conference, Weeds. British Crop Protection Council, Brighton, 3C-5, pp 297–302.

Holloway PJ, Western NM (2003) Tank-mix adjuvants and pesticide residues: some regulatory and quantitative aspects. Pestic Manag Sci 59:1237–1244.

Holloway PJ, Wong WWC, Partridge HJ, Seaman D, Perry RB (1992) Effects of some nonionic polyoxyethylene surfactants on uptake of ethirimol and diclobutrazol from suspension formulation applied to wheat leaves. Pestic Sci 34:109–118.

Holoman S Jr, Seymour KG (1983) Laboratory measurement of pesticide vapor losses. In: Kaneko TM, Akesson NB (eds) Pesticide Formulation and Application Systems: Third Symposium. ASTM Special Technical Publication 828. American Society for Testing and Materials, Philadelphia, pp 42–51.

Huggenberger FH, Letey J, Farmer WJ (1973) Effect of two nonionic surfactants on adsorption and mobility of selected pesticides in a soil-system. Soil Sci Soc Am Proc 37:215–219.

Iglesias-Jiménez E, Sánchez-Martin MJ, Sánchez-Camazano M (1996) Pesticide adsorption in a soil-water system in the presence of surfactants. Chemosphere 32:1771–1782.

Iglesias-Jiménez E, Poveda E, Sánchez-Martin MJ, Sánchez-Camazano M (1997) Effect of nature of exogeneous organic matter on pesticide sorption by the soil. Arch Environ Contam Toxicol 33:117–124.

Isensee AR, Jones GE (1971) Absorption and translocation of root and foliage applied 2,4-dichlorophenol, 2,7-dichlorodibenzo-*p*-dioxin, and 2,3,7,8-tetrachlorodibenzo-*p*-dioxin. J Agric Food Chem 19:1210–1214.

Isoda T, Yamasaki M, Yano H, Harada S (1997) Kinetic study of the protolysis of benzoic acid derivatives in a micelle of dodecylammonium chloride. J Chem Soc Faraday Trans 93:449–452.

Jafvert CT, van Hoof PL, Heath JK (1994) Solubilization of non-polar compounds by non-ionic surfactant micelles. Water Res 28:1009–1017.

Jaiswal PV, Ijeri VS, Srivastava AK (2001) Voltammetric behavior of certain vitamins and their determination in surfactant media. Anal Sci 17(suppl):741–744.

Jaynes WF, Boyd SA (1991) Clay mineral type and organic compound sorption by hexadecyltrimethylammonium-exchanged clays. Soil Sci Soc Am J 55:43–48.

Jiang X, Yediler A, Yufang S, Sun T, Kettrup A (2005) Effect of linear alkylbenzene sulfonate (LAS) on the mineralization, metabolism and uptake of ^{14}C-phenanthrene in a model ecosystem (water-plant-air). Chemosphere 61:741–751.

Johnson S, Dureja P (2002) Effect of surfactants on persistence of azadirachtin-A (Neem based pesticide). J Environ Sci Health B37:75–80.

Jones-Hughes T, Turner A (2005) Sorption of ionic surfactants to estuarine sediment and their influence on the sequestration of phenanthrene. Environ Sci Technol 39:1688–1697.

Kahn MN (1995) Effects of anionic micelles on the intramolecular general base-catalyzed hydrolysis of phenyl and methyl salicylates. J Mol Catal A Chemical 102:93–101.

Kahn MN, Arifin Z (1996) Effects of cationic micelles on rates and activation parameters of intramolecular general base-catalyzed hydrolysis of ionized salicylate esters. J Colloid Interface Sci 180:9–14.

Kahn SU, Marriage PB, Hamill AS (1981) Effects of atrazine treatment of a corn field using different application methods, times and additives on the persistence of residues in soil and their uptake by oat plants. J Agric Food Chem 29:216–219.

Kaka JS, Hayton WL (1978) Temperature and surfactant dependence of accumulation of 4-aminoantipyrine and ethanol in fish. J Pharm Sci 67:1558–1563.

Katagi (1995) Partition of organophosphorus pesticides into lipid bilayer membranes. J Pestic Sci 20:65–74.

Katagi (2002) Abiotic hydrolysis of pesticides in the aquatic environment. Rev Environ Contam Toxicol 175:79–261.

Katagi T (2004) Photodegradation of pesticides on plant and soil surfaces. Rev Environ Contam Toxicol 182:1–195.

Katagi T (2006) Behavior of pesticides in water-sediment systems. Rev Environ Contam Toxicol 187:133–251.

Kaushik P, Kumar R, Jain I, Patanjali PK (2005) Solubilization of chlorpyrifos in non-ionic surfactant-water mixed micellar system. Pestic Res J 17:43–45.

Kile DE, Chiou CT (1989) Water solubility enhancements of DDT and trichlorobenzene by some surfactants below and above the critical micelle concentration. Environ Sci Technol 23:832–838.

Kim BJ, Im SS, Oh SG (2001) Investigation on the solubilization locus of anilide-HCl in SDS micelles with ^{1}H-NMR spectroscopy. Langmuir 17:565–566.

Kim JH, Papiernik SK, Farmer WJ, Gan J, Yates SR (2003) Effect of formulation on the behavior of 1,3-dichloropropene in soil. J Environ Qual 32:2223–2229.

Knepper TP, Petrovic M, de Voogt P (2003) Occurrence of surfactants in surface waters and freshwater sediments. I. Alkylphenol ethoxylates and their degradation products. In: Knepper TP, Barceló D, de Voogt P (eds) Comprehensive Analytical Chemistry, vol 40. Elsevier, New York, pp 675–693.

Knowles A (2001) Trends in pesticide formulation. AGROW Report DS-215. PJB Publications, Surrey.

Ko SO, Schlautman MA, Carraway ER (1998) Partitioning of hydrophobic organic compounds to sorbed surfactants. 1. Experimental studies. Environ Sci Technol 32:2769–2775.

Koren E (1972) Leaching of trifluralin and oryzalin in soil with three surfactants. Weed Sci 20:230–232.

Krogh KA, Halling-Sørensen B, Mogensen BB, Vejrup KV (2003) Environmental properties and effects of nonionic surfactant adjuvants in pesticides: a review. Chemosphere 50:871–901.

Kucharski M (2003) Influence of herbicide and adjuvant application on residues in soil and plant of sugar beet. J Plant Prot Res 43:225–232.

Kucharski M, Sadowski J, Rola H (2002) The influence of adjuvants on phenmedipham and desmedipham degradation in soil. In: Proceedings, 12th European Weed Research Society Symposium, Wageningen, pp 122–123.

Kulshrestha G (1987) Dissipation of herbicide butachlor from different formulations in direct seeded and transplanted rice crop. Pesticides (March) 20–24.

Kwan CY, Chu W (2005) Reaction mechanism of photoreduction of 2,4-dichlorophenoxyacetic acid in surfactant micelles. Ind Eng Chem Res 44:1645–1651.

Laha S, Luthy RG (1991) Inhibition of phenanthrene mineralization by nonionic surfactants in water-soil systems. Environ Sci Technol 25:1920–1930.

Laha S, Luthy RG (1992) Effects of nonionic surfactants on the solubilization and mineralization of phenanthrene in soil-water systems. Biotechnol Bioeng 40:1367–1380.

Lamont RE, Ducker WA (1998) Surface-induced transformations for surfactant aggregates. J Am Chem Soc 120:7602–7607.

Lamoureax GL, Rusness DG (1995) Quinclorac absorption, translocation, metabolism, and toxicity in leafy spruce (*Euphorbia esula*). Pestic Biochem Physiol 53:210–226.

Larson RA, Jafvert CT, Boscá F, Marley KA, Miller PL (2000) Effects of surfactants on reduction and photolysis (>290 nm) of nitroaromatic compounds. Environ Sci Technol 34:505–508.

Lee JF, Crum JR, Boyd SA (1989) Enhanced retention of organic contaminants by soils exchanged with organic cations. Environ Sci Technol 23:1365–1372.

Lee JF, Liao PM, Kuo CC, Yang HT, Chiou CT (2000) Influence of a nonionic surfactant (Triton X-100) on contaminant distribution between water and several soil solids. J Colloid Interface Sci 229:445–452.

Lee JF, Hsu MH, Chao HP, Huang HC, Wang SP (2004) The effect of surfactants on the distribution of organic compounds in the soil solid/water system. J Hazard Mater B114:123–130.

Lee JF, Chang YT, Chao HP, Huang HC, Hsu MH (2006) Organic compound distribution between nonionic surfactant solution and natural solids: applicability of a solution property parameter. J Hazard Mater B129:282–289.

Lee KH, de Mayo P (1980) Biphasic photochemistry: photochemical regioselectivity and critical micelle concentration determination. Photochem Photobiol 31:311–314.

Lei XG, Tang XD, Liu YC (1991) Microenvironmental control of photochemical reactions. 3. Additive effects on micellar structure and properties of TX-100. Langmuir 7:2872–2876.

Lei ZF, Ye CM, Wang XJ (2001) Hydrolysis kinetics of atrazine and influence factors. J Environ Sci 13:99–103.

Leung JW (1994) A fluorometric method to determine rain fastness, volatilization and photostability of glyphosate from glass slides, after application of Vision® with two adjuvants. J Environ Sci Health B29:341–363.

Li Y, Yediler A, Ou Z, Conrad I, Kettrup A (2001) Effects of a non-ionic surfactant (Tween 80) on the mineralization, metabolism and uptake of phenanthrene in wheat-solution-lava microcosm. Chemosphere 45:67–75.

Lichtenstein EP (1966) Increase of persistence and toxicity of parathion and diazinon in soils with detergents. J Econ Entomol 59:985–993.

Lichtenstein EP, Schulz KR (1970) Volatilization of insecticides from various substrates. J Agric Food Chem 18:814–818.

Lichtenstein EP, Fuhremann TW, Schulz KR, Skrentny RF (1967a) Effect of detergents and inorganic salts in water on the persistence and movement of insecticides in soils. J Econ Entomol 60:1714–1721.

Lichtenstein EP, Fuhremann TW, Scopes NEA, Skrentny RF (1967b) Translocation of insecticides from soils into pea plants. Effects of the detergent LAS on translocation and plant growth. J Agric Food Chem 15:864–869.

Liu G, Dai S, Qian Y, Gan Q (2003) Experimental study on effect of anion surfactant on degradation rate of aldicarb in soil. J Environ Sci Health B38:405–416.

Liu GG, Roy D, Rosen MJ (2000) A simple method to estimate the surfactant micelle–water distribution coefficients of aromatic hydrocarbons. Langmuir 16:3595–3605.

Liu ZQ (2004) Bentazone uptake into plant foliage as influenced by surfactants and carrier pH. Aust J Agric Res 55:967–971.

Liu Z, Edwards DA, Luthy RG (1992) Sorption of non-ionic surfactants onto soil. Water Res 26:1337–1345.

Locke MA, Reddy KN, Gaston LA, Zablotowicz RM (2002) Adjuvant modification of herbicide interactions in aqueous soil suspensions. Soil Sci 167:444–452.

Loh W, Volpe PLO (1992) Interaction of alkyl *p*-hydroxybenzoates with micelles: evaluation of their partition coefficients and ionization constants. J Colloid Interface Sci 154:369–377.

Lopes A, Maganita AL, Pina FS, Melo E, Wamhoff H (1992) Multiequilibria of 2-(2®-furanyl)-1*H*-benzimidazole neutral and protonated forms in the presence of amphiphilic aggregates. Environ Sci Technol 26:2448–2453.

Lopes A, de Melo JS, Martins AJ, Maganita AL, Pina FS, Wamhoff H, Melo E (1995) Partition of pesticides of the coumarin family between water and amphiphilic aggregations. Environ Sci Technol 29:562–570.

Luiz M, Gutiérrez MI, Bocco G, Bertolotti SG, Garcia NA (1992) Sensitized photooxidation of mononitrophenols in micellar media. Toxicol Environ Chem 35:115–123.

Mackay RA (1981) Chemical reactions in microemulsions. Adv Colloid Interface Sci 15:131–156.

Mackay RA, Hermansky C (1981) Phosphate ester–nucleophile reactions in oil-in-water microemulsions. J Phys Chem 85:739–744.

Maguire RJ (1991) Kinetics of pesticide volatilization from the surface of water. J Agric Food Chem 39:1674–1678.

Makkar RS, Rockne KJ (2003) Comparison of synthetic surfactants and biosurfactants in enhancing biodegradation of polycyclic aromatic hydrocarbons. Environ Toxicol Chem 22:2280–2292.

Manohar C, Rao URK, Valaulikar BS, Iyer RM (1986) On the origin of viscoelasticity in micellar solutions of cetyltrimethylammonium bromide and sodium salicylate. J Chem Soc Chem Commun 379–381.

Mansager ER, Still GG, Frear DS (1979) Fate of [^{14}C]diflubenzuron on cotton and in soil. Pestic Biochem Physiol 12:172–182.

Marchesi JR, Russell NJ, White GF, House WA (1991) Effects of surfactant adsorption and biodegradability on the distribution of bacteria between sediments and water in a freshwater microcosm. Appl Environ Microbiol 57:2507–2513.

Margoum C, Malessard C, Gouy V (2006) Investigation of various physicochemical and environmental parameter influence on pesticide sorption to ditch bed substratum by means of experimental design. Chemosphere 63:1835–1841.

Marshall DB, Pree DJ (1993) Factors affecting toxicity of propargite to the European red mite (*Acari*: *Tetranychidae*). Hortic Entomol 86:854–859.

Martínez MJ, Farsaoui K, de Prado R (2004) Weed control and persistence of two oxyfluorfen formulations in olive groves under non tillage conditions. Commun Appl Biol Sci Ghent Univ 69/3:77–82.

Martins JMF, Mermoud A (1999) Transport of rimsulfuron and its metabolites in soil columns. Chemosphere 38:601–616.

Mata JP, Aswal VK, Hassan PA, Bahadur P (2006) A phenol-induced structural transition in aqueous cetyltrimethylammonium bromide solution. J Colloid Interface Sci 299:910–915.

Mata-Sandoval JC, Karns J, Torrents A (2000) Effects of rhamnolipids produced by *Pseudomonas aeruginosa* UG2 on the solubilization of pesticides. Environ Sci Technol 34:4923–4930.

Mata-Sandoval JC, Karns J, Torrents A (2001) Influence of rhamnolipids and Triton X-100 on the biodegradation of three pesticides in aqueous phase and soil slurries. J Agric Food Chem 49:3296–3303.

Mata-Sandoval JC, Karns J, Torrents A (2002) Influence of rhamnolipids and Triton X-100 on the desorption of pesticides from soil. Environ Sci Technol 36:4669–4675.

Matondo H, de Savignac A, Bergon M, Calmon JP, Lattes A (1990) Kinetics of the hydrolysis of the potentially pesticidal *N*-(4-pyridyl)carbamates in micellar solutions. J Agric Food Chem 38:1106–1109.

McInnes D, Harker KN, Blackshaw RE, van den Born WH (1992) The influence of ultraviolet light on the phytotoxicity of sethoxydim tank mixtures with various adjuvants. In: Foy CL (ed) Adjuvants for Agrichemicals. CRC Press, Boca Raton, pp 205–213.

McWhorter CG (1985) The physiological effects of adjuvants on plants. In: Duke DO (ed) Weed Physiology, vol II. Herbicide Physiology. CRC Press, Boca Raton, pp 141–158.

Mekkaoui M, El Azzouzi M, Bouhaouss A, Ferhat M, Dahchour A, Guittonneau S, Meallier P (2000) Factors influencing imazapyr herbicide photolysis in water. Fresenius Environ Bull 9:783–790.

Menger FM, Portnoy CE (1967) On the chemistry of reactions proceeding inside molecular aggregates. J Am Chem Soc 89:4698–4703.

Mielbrecht EE, Wolfe MF, Tjeerdema RS, Sowby ML (2005) Influence of a dispersant on the bioaccumulation of phenanthrene by topsmelt (*Atherinops affinis*). Ecotox Environ Safety 61:44–52.

Miller DE, Shaw FR, Smith CT (1969) The comparative residual life of two formulations of Imidan® on alfalfa. J Econ Entomol 62:720–721.

Miller RM (1995) Surfactant-enhanced bioavailability of slightly soluble organic compounds. In: Skipper HD, Turco RF (eds) Bioremediation: Science and Applications. Soil Sci Soc Special Publ 43. Soil Science Society of America, Madison, pp 33–54.

Mingelgrin U, Yaron B (1973) Conversion of some organophosphorus insecticides on adsorbing surfaces as affected by formulation. Bull Environ Contam Toxicol 10:285–290.

Mirgorodskaya AB, Kudryavtseva LA (2002) Nucleophilic substitution in carboxylic esters in oil-in-water microemulsions. Russ J Gen Chem 72:1261–1265.

Mishael YG, Undabeytia T, Rytwo G, Papahadjopoulos-Sternberg B, Rubin B, Nir S (2002) Sulfometuron incorporation in cationic micelles adsorbed on montmorillonite. J Agric Food Chem 50:2856–2863.

Mishael YG, Undabeytia T, Rabinovitz O, Rubin B, Nir S (2003) Sulfosulfuron incorporated in micelles adsorbed on montmorillonite for slow release formulations. J Agric Food Chem 51:2253–2259.

Mishra P, Mishra BK, Behera GB (1992) Hydrolysis of Schiff bases. 2. Intramolecular catalysis of an *ortho*-hydroxy group in nonionic surfactant systems. Int J Chem Kinet 24:593–618.

Mollett KJ, O'Connor CJ (1976) Hydrolysis of phenylureas. Part III. Micellar effects on the solubilization and decomposition of 4-methyl- and 4-nitro-phenyl urea. J Chem Soc Perkin Trans 2:369–374.

Monkiedje A, Spiteller M, Bester K (2003) Degradation of racemic and enantiopure metalaxyl in tropical and temperate soils. Environ Sci Technol 37:707–712.

Montemurro N, Grieco F, Lacertosa G, Visconti A (2002) Chlorpyrifos decline curves and residue levels from different commercial formulations applied to oranges. J Agric Food Chem 50:5975–5980.

Moreno MJ, Herrmann C, Ivo R, Lourtie IMG, Wamhoff H, Melo E (1995) Consequences of the partition between water and molecular aggregates on the photodegradation pattern and kinetics of a pesticide of the coumarin family. Environ Sci Technol 29:136–141.

Mukerjee P, Banerjee K (1964) A study of the surface pH of micelles using solubilized indicator dyes. J Phys Chem 68:3567–3574.
Mukherjee I, Gopal M (1999) Behavior of different formulations of lindane on chickpea. Bull Environ Contam Toxicol 63:467–472.
Mulder H, Wassink GR, Breure AM, van Andel JG, Rulkens WH (1998) Effect of nonionic surfactants on naphthalene dissolution and biodegradation. Biotechnol Bioeng 60:397–407.
Müller MT, Zehnder AJB, Escher BI (1999) Membrane toxicity of linear alcohol ethoxylates. Environ Toxicol Chem 18:2767–2774.
Nakamura Y (1988) Regioselective photodimerization of cinnamic acid in a micellar solution. J Chem Soc Chem Commun 477–478.
Nayyar SP, Sabatini DA, Harwell JH (1994) Surfactant adsolubilization and modified admicellar sorption of nonpolar, polar and ionizable organic contaminants. Environ Sci Technol 28:1874–1881.
Ngim KK, Crosby DG (2001) Abiotic processes influencing fipronil and desthiofipronil dissipation in California, USA, rice fields. Environ Toxicol Chem 20:972–977.
Nielsen CM, Steele KD, Forster WA, Zabkiewicz JA (2005) Influence of dose and molecular weight on foliar mass uptake of surfactant. N Z Plant Prot 58:174–178.
Nome F, Schwingel EW, Ionescu LG (1980) Micellar effects on the base-catalyzed oxidative cleavage of a carbon-carbon bond in 1,1-bis(*p*-chlorophenyl)-2,2,2-trichloroethanol. J Org Chem 45:705–710.
Nome F, Rublra AF, Franco C, Ioneecu LG (1982) Limitation of the pseudophase model of micellar catalysis. The dehydrochlorination of 1,1,1-trichloro-2,2-bis(*p*-chlorophenyl)ethane and some of its derivatives. J Phys Chem 86:1881–1885.
Nutahara M, Murai T (1984) Accelerating effects of natural unsaturated fatty acids on photodegradation of chinomethionat (Morestan®). J Pestic Sci 9:667–674.
O'Connor CJ, Tan AL (1980) Micellar catalyzed hydrolysis of amides. 4-Nitroacetanilide in hexadecyltrimethylammonium bromide. Aust J Chem 33:747–755.
OECD (1996) OECD Guideline for the testing of chemicals. Bioconcentration: flow-through fish test, no. 305. OECD, Brussels.
OECD (2000) OECD Guideline for the testing of chemicals. Adsorption–desorption using a batch equilibrium method, no. 106. OECD, Brussels.
OECD (2002a) OECD Guideline for the testing of chemicals. Aerobic and anaerobic transformation in soil, no. 307. OECD, Brussels.
OECD (2002b) OECD Guideline for the testing of chemicals. Aerobic and anaerobic transformation in aquatic sediment systems, no. 308. OECD, Brussels.
OECD (2004) OECD Guideline for the testing of chemicals. Hydrolysis as a function of pH, no. 111. OECD, Brussels.
OECD (2007) OECD Guideline for the testing of chemicals. Metabolism in Crops, no. 501. OECD, Brussels.
Orr GL, Hogan ME (1985) UV-A photooxidation of β-carotene in Triton X-100 micelles by nitrodiphenyl ether herbicides. J Agric Food Chem 33:968–972.
Otero C, Rodenas E (1986) Influence of *n*-butyl and *n*-hexyl alcohols in the dehydrohalogenation of DDT in cationic micelles of *N*-cetyl-*N*,*N*,*N*-trimethylammonium bromide, chloride, and hydroxide. J Phys Chem 90:5771–5775.
Paria S, Yuet PK (2006) Solubilization of naphthalene by pure and mixed surfactants. Ind Eng Chem Res 45:3552–3558.
Park JW, Jaffé PR (1993) Partitioning of three nonionic organic compounds between adsorbed surfactants, micelles and water. Environ Sci Technol 27:2559–2565.
Park SS, Park JW, Uchrin C, Cheney MA (2002) A micelle inhibition model for the bioavailability of polycyclic aromatic hydrocarbons in aquatic systems. Environ Toxicol Chem 21:2737–2741.
Patel JM, Wurster DE (1991) Catalysis of carbaryl hydrolysis in micellar solutions of cetyltrimethyl-ammonium bromide. Pharm Res 8:1155–1158.
Paul BK, Moulik SP (2001) Uses and application of microemulsions. Curr Sci 80:990–1001.

Pedersen JA, Yeager MA, Suffet IHM (2003) Xenobiotic organic compounds in runoff from fields irrigated with treated wastewater. J Agric Food Chem 51:1360–1372.
Pennell KD, Karagunduz A, Young MH (2003) Impacts of surfactant adjuvants on pesticide availability and transport in soils. In: Gan JJ, Zhu PC, Aust SD, Lemley AT (eds) Pesticide Decontamination and Detoxification. ACS Symp Ser 863, American Chemical Society, Washington, DC, pp 231–245.
Petersen PJ, Haderlie LC, Hoefer RH, McAllister RS (1985) Dicamba absorption and translocation as influenced by formulation and surfactant. Weed Sci 33:717–720.
Piekarski D, Potier P, Giannotti C (1990) Study of the mode of action of some nitrophenyl ethers. Free Radical Res Commun 10:85–93.
Piszkiewicz D (1977) Positive cooperativity in micelle-catalyzed reactions. J Am Chem Soc 99:1550–1557.
Polubesova T, Nir S, Rabinovitz O, Borisover M, Rubin B (2003) Sulfentrazone adsorbed on micelle–montmorillonite complexes for slow release in soil. J Agric Food Chem 51:3410–3414.
Porras-Rodriguez M, Talens-Alesson FI (1999) Removal of 2,4-dichlorophenoxyacetic acid from water by adsorptive micellar flocculation. Environ Sci Technol 33:3206–3209.
Pourreza N, Rastegarzadeh S (2005) Spectrophotometric determination of the dissociation constant of 5-(*p*-dimethylaminobenzylidene)rhodamine in micellar media. J Chem Eng Data 50:206–210.
Pree DJ, Butler KP, Kimball ER, Stewart DKR (1976) Persistence of foliar residues of dimethoate and azinphos-methyl and their toxicity to the apple maggot. J Econ Entomol 69:473–478.
Purnanand, Danikhel RK (1985) Micellar catalysis on supernucleophile (HOO^-) assisted hydrolysis. J Surface Sci Technol 1:69–72.
Putnam RA, Nelson JO, Clark JM (2003) The persistence and degradation of chlorotalonil and chlorpyrifos in a cranberry bog. J Agric Food Chem 51:170–176.
Que Hee SS, Sutherland RG (1975) Factors affecting the volatility of DDT, dieldrin and dimethylamine salt of (2,4-dichlorophenoxy)acetic acid (2,4-D) from leaf and glass surfaces. Bull Environ Contam Toxicol 13:284–290.
Que Hee SS, Paine SH, Sutherland RG (1979) Photodecomposition of a formulated mixed butyl ester of 2,4-dichlorophenoxyacetic acid in aqueous and hexane solutions. J Agric Food Chem 27:79–82.
Quintero JC, Moreira MT, Feijoo G, Lema JM (2005) Effect of surfactants on the soil desorption of hexachlorohexane (HCH) isomers and their anaerobic biodegradation. J Chem Tech Biotechnol 80:1005–1015.
Ramamurthy V (1986) Organic photochemistry in organized media. Tetrahedron 42:5753–5839.
Rawlins TE, Booth JA (1968) Tween 20 as an adjuvant for systemic soil fungicides for *Verticillium* in cotton. Plant Dis Rep 52:944–945.
Reeves B (1993) The use of a 'sticker' type of adjuvant with insecticides. Pestic Sci 37:206–207.
Rodriguez-Cruz MS, Sánchez-Martin MJ, Sánchez-Camazano M (2001) Degradation of linuron in soils as influenced by different organic amendments and surfactants. Proceedings, BCPC Symposium: Pesticide Behavior in Soils and Water. British Crop Protection Council, Brighton, vol 78, pp 139–144.
Rodriguez-Cruz MS, Sánchez-Martin MJ, Sánchez-Camazano M (2005) A comparative study of adsorption of an anionic and a non-ionic surfactant by soils based on physicochemical and mineralogical properties of soils. Chemosphere 61:56–64.
Rodriguez-Cruz MS, Sánchez-Martin MJ, Sánchez-Camazano M (2006) Surfactant-enhanced desorption of atrazine and linuron residues as affected by aging of herbicides in soil. Arch Environ Contam Toxicol 50:128–137.
Rodríguez-Gonzalo E, Sánchez-Martín MJ, Sánchez-Camazano M (1993) Adsorption of parathion and paraoxon by modified montmorillonite. Toxicol Environ Chem 37:157–163.
Romsted LR, Cordes EH (1968) Secondary valence force catalysis. VII. Catalysis of hydrolysis of *p*-nitrophenyl hexanoate by micelle-forming cationic detergents. J Am Chem Soc 90:4404–4409.

Rouse JD, Sabatini DA, Suflita JM, Harwell JH (1994) Influence of surfactants on microbial degradation of organic compounds. Crit Rev Environ Sci Technol 24:325–370.
Rüdel H (1997) Volatilization of pesticides from soil and plant surfaces. Chemosphere 35:143–152.
Ruzo LO (1983) Involvement of oxygen in the photoreactions of cypermethrin and other halogenated pyrethroids. J Agric Food Chem 31:1113–1115.
Ruzo LO, Krishnamurthy VV, Casida JE, Gohre K (1987) Pyrethroid photochemistry: influence of the chloro(trifluoromethyl)vinyl substituent in cyhalothrin. J Agric Food Chem 35:879–883.
Saha SK, Tiwari PK, Dogra SK (1994) Prototropic equilibrium of some benzimidazoles in anionic and nonionic micelles. J Phys Chem 98:5953–5955.
Sánchez L, Romero E, Peña A (2003a) Ability of biosolids and a cationic surfactant to modify methidathion leaching. Modelling with pescol. Chemosphere 53:843–850.
Sánchez L, Romero E, Sánchez-Rasero F, Gios G, Peña A (2003b) Enhanced soil sorption of methidathion using sewage sludge and surfactants. Pestic Manag Sci 59:857–864.
Sánchez L, Romero E, Peña A (2005) Photostability of methidathion in wet soil amended with biosolid and a surfactant under solar irradiation. Chemosphere 59:969–976.
Sánchez M, Aranda FJ, Espuny MJ, Marqués A, Teruel JA, Manresa Á, Ortiz A (2007) Aggregation behaviour of a dirhamnolipid biosurfactant secreted by *Pseudomonas aeruginosa* in aqueous media. J Colloid Interface Sci 307:246–253.
Sánchez-Camazano M, Sánchez-Martin MJ (1994) Organo-clays as adsorbents for azinphos-methyl and dichlorvos in aqueous medium. Water Air Soil Pollut 74:19–28.
Sánchez-Camazano M, Arienzo M, Sánchez-Martin MJ, Crisanto T (1995) Effect of different surfactants on the mobility of selected non-ionic pesticides in soil. Chemosphere 31:3793–3801.
Sánchez-Camazano M, Sánchez-Martin MJ, Delgado-Pascual R (2000a) Adsorption and mobility of linuron in soils as influenced by soil properties, organic amendments, and surfactants. J Agric Food Chem 48:3018–3026.
Sánchez-Camazano M, Sánchez-Martin MJ, Rodriguez-Cruz MS (2000b) Sodium dodecyl sulfate- enhanced desorption of atrazine: effect of surfactant and of organic matter content of soil. Chemosphere 41:1301–1305.
Sánchez-Camazano M, Rodriguez-Cruz MS, Sánchez-Martin MJ (2003) Evaluation of component characteristics of soil- and surfactant-herbicide system that affect enhanced desorption of linuron and atrazine preadsorbed by soils. Environ Sci Technol 37:2758–2766.
Sanderson H, Dyer SD, Price BB, Nielsen AM, van Compernolle R, Selby M, Stanton K, Evans A, Ciarlo M, Sedlak R (2006) Occurrence and weight-of-evidence risk assessment of alkylsulfates, alkylethoxykates, and linear alkylbenzenesulfonates (LAS) in river and sediments. Sci Total Environ 368:695–712.
Sands R, Bachelard EP (1973) Uptake of picloram by *Eucalypt* leaf discs. I. Effect of surfactants and nature of the leaf surfaces. New Phytol 72:69–86.
Santoro A, Scopa A, Bufo SA, Mansour M, Mountacer H (2000) Photodegradation of the triazole fungicide hexaconazole. Bull Environ Contam Toxicol 64:475–480.
Sarkar MA, Roy S, Kole RK, Chowdhury A (2001) Persistence and metabolism of imidacloprid in different soils of West Bengal. Pestic Manag Sci 57:598–602.
Sarpal RS, Dogra SK (1993) Prototropic equilibrium of some aromatic amines in micellar solutions of sodium dodecyl sulphate. J Photochem Photobiol A Chem 69:329–335.
Sassman SA, Lee LS, Bischoff M, Tureo RF (2004) Assessing *N,N'*-dibutylurea (DBU) formation in soils after application of *n*-butylisocyanate and benlate fungicides. J Agric Food Chem 52:747–754.
Scaiano JC, Abuin EB, Stewart LC (1982) Photochemistry of benzophenone in micelles. Formation and decay of radical pairs. J Am Chem Soc 104:5673–5679.
Schanze KS, Shin DM, Whitten DG (1985) Micelle and vesicle solubilization sites. Determination of micropolarity and microviscosity using photophysics of a dipolar olefin. J Am Chem Soc 107:507–509.

Scheunert I, Korte F (1985) Interactions in the fate of chemicals in terrestrial systems. Ecotoxicol Environ Saf 9:385–391.
Schönherr J, Bauer H (1992) Analysis of effects of surfactants on permeability of plant cuticles. In: Foy CL (ed) Adjuvants for Agrichemicals. CRC Press, Boca Raton, pp 17–35.
Schönherr J, Riederer M (1989) Foliar penetration and accumulation of organic chemicals in plant cuticles. Rev Environ Contam Toxicol 108:1–70.
Schönherr J, Riederer M, Schreiber L, Bauer H (1991) Foliar uptake of pesticides and its activation by adjuvants: theories and methods for optimization. In: Frehse H (ed) Pesticide Chemistry. VCH Verlag, Weinheim, pp 237–253.
Schrap SM (1991) Bioavailability of organic chemicals in the aquatic environment. Comp Biochem Physiol 100C:13–16.
Schreiber L (1994) A mechanistic approach towards effects of nonionic surfactants on mobility of pesticides in reconstituted cuticular wax of barley leaves. Med Fac Landbouww Univ Gent 59/3b:1409–1414.
Schreiber L, Schönherr J (1993) Mobilities of organic compounds in reconstituted cuticular wax of barley leaves: determination of diffusion coefficients. Pestic Sci 38:353–361.
Schulz KR, Lichtenstein EP, Liang TT, Fuhremann TW (1970) Persistence and degradation of azinphos-methyl in soils, as affected by formulation and mode of application. J Econ Entomol 63:432–438.
Scrano L, Bufo SA, Perucci P, Meallier P, Mansour M (1999) Photolysis and hydrolysis of rimsulfuron. Pestic Sci 55:955–961.
Seguchi K (1979) Effects of surfactants on the visible spectra and acidity of substituted nitrophenols. Yukagaku 28:20–25.
Shang C, Rice JA (2007) Investigation of humate-cetyltrimethylammonium complexes by small-angle X-ray scattering. J Colloid Interface Sci 305:57–61.
Sharma MP, van den Born WH, Friesen HA, McBeath DK (1976) Penetration, translocation and metabolism of ^{14}C-difenzoquat in wild oat and barley. Weed Sci 24:379–384.
Sharma SR, Singh RP, Ahmad SR (1985) Effect of different saline, alkaline salts, fertilizers and surfactants on the movement of some phosphorus-containing pesticides in soils. Ecotoxicol Environ Saf 10:339–350.
Sheng JJ, Kasin NA, Chandrasekharan R, Amidon GL (2006) Solubilization and dissolution of insoluble weak acid, ketoprofen: effects of pH combined with surfactant. Eur J Pharm 29:306–314.
Shi Z, Sigman ME, Ghosh MM, Dabestani R (1997) Photolysis of 2-chlorophenol dissolved in surfactant solutions. Environ Sci Technol 31:3581–3587.
Shi Z, Ghosh MM, Robinson KG (2000) Surfactant-enhanced photolysis of polychlorinated biphenyls (PCBs): a solution for remediation of PCB contaminated soils. In: Wise DL, Trantolo DJ, Eichon EJ, Inyang HI, Stottmeister V (eds) Remediation Engineering of Contaminated Soils. Environmental Science and Pollution Control Series, vol 23. Dekker, New York, pp 523–540.
Sigman ME, Schuler PF, Ghosh MM, Dabestani RT (1998) Mechanism of pyrene photochemical oxidation in aqueous and surfactant solutions. Environ Sci Technol 32:3980–3985.
Šimá®ová E, Shi T, Schönherr J, Schreiber L (2005) Sorption in reconstituted waxes of homologous series of alcohol ethoxylates and *n*-alkyl esters and their effects on the mobility of 2, 4-dichlorophenoxybutyric acid. Pestic Manag Sci 61:383–389.
Singh AK, Raghuraman TS (1985) Photorearrangement of phenyl cinnamates under micellar environment. Tetrahedron Lett 26:4125–4128.
Singh AK, Raghuraman TS (1986) Photobehaviour of *N*-aryl amides in micelle. Synth Commun 16:485–490.
Singh AK, Sonar SM (1985) Photorearrangement of aryl esters in micellar medium. Synth Commun 15:1113–1122.
Singh N (2006) Reduced downward mobility of metolachlor and metribuzin from surfactant-modified clays. J Environ Sci Health 41B:17–29.

Singh N, Megharaj M, Gates WP, Churchman GJ, Anderson J, Kookana RS, Naidu R, Chen Z, Slade PG, Sethunathan N (2003) Bioavailability of an organophosphorus pesticide, fenamiphos, sorbed on an organo clay. J Agric Food Chem 51:2653–2658.

Singh N, Megharaj M, Gates WP, Churchman J, Kookana RS, Naidu R, Sethunathan N (2004) Sorption-desorption of fenamiphos in surfactant-modified clays. Bull Environ Contam Toxicol 72:276–282.

Singh RP, Kumar R (2000) Evaluation of the effect of surfactants on the movement of pesticides in soils using a soil thin-layer chromatography technique. Soil Sed Contam 9:407–423.

Singh RP, Rawat JP, Kumar R (2000) Effect of cationic, non-ionic and anionic surfactants on the adsorption of carbofuran on three different types of Indian soils. Adv Sci Technol 18:333–346.

Singh M, Tan S, Sharma SD (2002a) Effects of adjuvants on wetting and water infiltration of soils. Bull Environ Contam Toxicol 68:692–698.

Singh M, Tan S, Sharma SD (2002b) Leaching and sorption of norflurazon in soils as affected by cationic surfactants. Bull Environ Contam Toxicol 68:901–907.

Smith AM, Chow PNP (1990) The influence of agral 90 surfactant on the activity of imazamethabenz in wild oats (*Avena fatua L.*). Weed Res 30:355–362.

Smith CN, Nalewaja JD (1972) Uptake and translocation of foliarly-applied atrazine. Weed Sci 20:36–40.

Smith GA, Christian SD, Tucker EE, Scamehorn JF (1987) Group contribution model for predicting the solubilization of organic solutes by surfactant micelles. Langmuir 3:598–599.

Smith JH, Bomberger DC Jr, Haynes DL (1980) Prediction of the volatilization rates of high-volatility chemicals from natural water bodies. Environ Sci Technol 14:1332–1337.

Stadler E, Zanette D, Rezende MC, Nome F (1984) Kinetic behavior of cetyltrimethylammonium hydroxide. The dehydrochlorination of 1,1,1-trichloro-2,2-bis(*p*-chlorophenyl)ethane and some of its derivatives. J Phys Chem 88:1892–1896.

Stanger CE, Vargas TC (1984) A photodecomposition study of napropamide. Proc West Soc Weed Sci 37:221–225.

Staudinger J, Roberts PV (1996) A critical review of Henry's law constants for environmental allpications. Crit Rev Environ Sci Technol 26:205–297.

Steurbaut W (1994) Influence of surfactants on the behavior of pesticides in soil. In: Proceedings of 5th International Workshop, Environmental Behavior of Pestic cides, Regulatory Aspects, Rixensart, Belgium. Section II: Sorption, pp 193–197.

Stevens PJG (1993) Organosilicone surfactants as adjuvants for agrochemicals. Pestic Sci 38:103–122.

Stevens PJG, Baker EA, Anderson NH (1988) Factors affecting the foliar absorption and redistribution of pesticides. 2. Physicochemical properties of the active ingredient and the role of surfactant. Pestic Sci 24:31–53.

Stevens PJG, Gaskin RE, Hong SO, Zabkiewicz JA (1992) Pathways and mechanisms of foliar uptake as influenced by surfactants. In: Foy CL (ed) Adjuvants for Agrichemicals. CRC Press, Boca Raton, pp 385–398.

Stock D, Holloway PJ, Grayson BT, Whitehouse P (1993) Development of a predictive uptake model to rationalize selection of polyoxyethylene surfactant adjuvants for foliage-applied agrochemicals. Pestic Sci 37:233–245.

Stolzenberg GE, Olson PA, Zaylskie RG, Mansager ER (1982) Behavior and fate of ethoxylated alkyphenol nonionic surfactant in barley plants. J Agric Food Chem 30:637–644.

Stolzenberg GE, Olson PA, Tanaka FS, Mansager ER, Lamoureux CH (1984) Identification of glucoylated conjugates and oxygenated metabolites of nonionic surfactants in barley and rice leaf tissues. In: Scher HB (ed) Advances in Pesticide Formulation Technology. ACS Symp Ser 254. American Chemical Society, Washington, DC, pp 207–218.

Sugimura Y, Takeno T (1985) Behaviour of polyoxyethylene sorbitan ^{14}C-monooleate in tobacco and kidney bean leaves. J Pestic Sci 10:233–239.

Sun S, Inskeep WP (1995) Sorption of nonionic organic compounds in soil-water systems containing micelle-forming surfactant. Environ Sci Technol 29:903–913.

Sundaram A (1985) A gravimetric method for determining the relative volatilities of non-aqueous pesticide formulations and spray diluents. Pestic Sci 16:397–403.

Sundaram A (1987) Intermolecular interactions in aqueous pesticide formulations from viscosity—temperature, viscosity–volatility, and viscosity—shear rate relationships. In: van der Hooven DIB, Spicer LD (eds) Pesticide Formulations and Application Systems. ASTM Special Technical Publication 943. American Society for Testing and Materials, Philadelphia, pp 48–67.

Sundaram A (1995) Physical properties and evaporation characteristics of nonaqueous insecticide formulations, spray diluents and adjuvant/co-solvent mixtures. J Environ Sci Health B30: 113–138.

Sundaram A, Leung JW (1986) A simple method to determine relative volatilities of aqueous formulations of pesticides. J Environ Sci Health B21:165–190.

Sundaram KMS (1986) A comparative evaluation of dislodgeable and penetrated residues, and persistence characteristics of aminocarb and fenitrothion, following application of several formulations onto conifer trees. J Environ Sci Health B21:539–560.

Sundaram KMS (1997) Effect of additives in the neem formulation on deposition, volatilization and persistence of azadirachtin in spruce foliage. J Environ Sci Health B32:523–544.

Suratkar V, Mahapatra S (2000) Solubilization site of organic perfume molecules in sodium dodecyl sulfate micelles: new insights from proton NMR studies. J Colloid Interface Sci 225:32–38.

Swarcewicz MK, Muliñski Z, Zbie® I (1998) Influence of spray adjuvants on the behaviour of trifluralin in the soil. Bull Environ Contam Toxicol 60:569–576.

Szeto SY, Burlinson NE, Rahe JE, Oloffs PC (1989) Persistence of the fungicide vinclozolin on pea leaves under laboratory conditions. J Agric Food Chem 37:529–534.

Tan S, Singh M (1995) Effects of cationic surfactants on leaching of bromacil and norflurazon. Bull Environ Contam Toxicol 55:359–365.

Tanaka FS (1989) Surfactant and herbicide interactions during photolysis with ultraviolet light. In: Chow PNP, Grant CA, Hinshalwood AM, Simundsson E (eds) Adjuvants and Agrochemicals, vol 2. CRC Press, Boca Raton, pp 15–24.

Tanaka FS, Wien RG, Mansager ER (1979) Effect of nonionic surfactants on the photochemistry of 3-(4-chlorphenyl)-1,1-urea in aqueous solution. J Agric Food Chem 27:774–779.

Tanaka FS, Wien RG, Mansager ER (1981) Survey for surfactant effects on the photodegradation of herbicides in aqueous media. J Agric Food Chem 29:227–230.

Tanaka FS, Wien RG, Hoffer BL (1982) Investigation of the mechanism and pathway of biphenyl formation in the photolysis of monuron. J Agric Food Chem 30:957–963.

Tanaka FS, Wien RG, Hoffer BL (1986) Photosensitized degradation of a homogeneous nonionic surfactant: hexaethoxylated 2,6,8-trimethyl-4-nonanol. J Agric Food Chem 34:547–551.

Tanaka FS, Wien RG, Zaylskie RG (1991) Photolytic degradation of a homogeneous Triton X nonionic surfactant: nonaethoxylated *p*-(1,1,3,3-tetramethylbutyl)phenol. J Agric Food Chem 39:2046–2052.

Tao QH, Wang DS, Tang HX (2006) Effect of surfactants at low concentrations on the sorption of atrazine by natural sediment. Water Environ Res 78:653–660.

Tas¸cioğlu S (1996) Micellar solutions as reaction media. Tetrahedron 52:11113–11152.

Tee OS, Federtchenko AA (1997) Transition state stabilization by micelles: the hydrolysis of *p*-nitrophenyl alkanoates in cetyltrimethylammonium bromide micelles. Can J Chem 75:1434–1438.

Telscher MJH, Schuller U, Schmidt B, Schäffer A (2005) Occurrence of a nitro metabolite of a defined nonylphenol isomer in soil/sewage sludge mixture. Environ Sci Technol 39:7896–7900.

Theng BKG, Newman RH, Whitton JS (1998) Characterization of an alkylammonium-montmorillonite- phenanthrene intercalation complex by carbon-13 nuclear magnetic resonance spectroscopy. Clay Miner 33:221–229.

Theng BKG, Aislabie J, Fraser R (2001) Bioavailability of phenanthrene intercalated into an alkylammonium-montmorillonite clay. Soil Biol Biochem 33:845–848.

Thomas JK (1977) Effect of structure and charge on radiation-induced reactions in micellar systems. Acc Chem Res 10:133–138.
Thomas JK (1980) Radiation-induced reactions in organized assemblies. Chem Rev 80:283–299.
Thomas SM, Harrison SK (1990) Surfactant-altered rates of chlorimuron and metsulfuron photolysis in sunlight. Weed Sci 38:602–606.
Thompson WM, Nissen SJ (2000) Absorption and fate of carfentrazone-ethyl in *Zea mays*, *Glycine max*, and *Abutilon theophrasti*. Weed Sci 48:15–19.
Tiehm A, Stieber M, Werner P, Frimmel FH (1997) Surfactant-enhanced mobilization and biodegradation of polycyclic aromatic hydrocarbons in manufactured gas plant soil. Environ Sci Technol 31:2570–2576.
Tolls J, Kloepper-Sams P, Sijm DTHM (1994) Surfactant bioconcentration: a critical review. Chemosphere 29:693–717.
Turro NJ, Cherry WR (1978) Photoreaction in detergent solution. Enhancement of regioselectivity resulting from the reduced dimensionality of substrates sequestered in a micelle. J Am Chem Soc 100:7431–7432.
Turro NJ, Liu KC, Chow MF (1977) Solvent sensitivity of type II photoreaction of ketones as a device to probe solute location in micelles. Photochem Photobiol 26:413–415.
Turro NJ, Zimmt MB, Lei XG, Gould IR, Nitsche KS, Cha Y (1987) Additive effects on the CIDNP, cage effect and exit rate of micellized radical pairs. J Phys Chem 91:4544–4548.
Turro NJ, Grätzel M, Braun AM (1980) Photophysical and photochemical processes in micellar systems. Angew Chem Int Ed Engl 19:675–696.
Urano K, Saito M, Murata C (1984) Adsorption of surfactants on sediments. Chemosphere 13:293–300.
Valoras N, Letey J, Osborn J (1974) Uptake and translocation of a nonionic surfactant by barley. Agronomy J 66:436–438.
Valsaraj KT (1992) Separation of hydrophobic organic compounds from waste water using surfactant aggregates on alumina particles. Water Sci Technol 26:1213–1220.
Valsaraj KT, Gupta A (1988) Partitioning of chloromethanes between aqueous and surfactant micellar phases. Water Res 22:1173–1183.
Valsaraj KT, Thibodeaux LJ (1989) Relationships between micelle-water and octanol-water partition coefficients for hydrophobic organics of environmental interest. Water Res 23:183–189.
Van Eden CM, Liebenberg W, du Preez JL, de Villiers MM (2004) Solvent and surfactant enhanced solubilization, stabilization and degradation of amitraz. J Environ Sci Health B39:33–51.
Van Ginkel CG (1996) Complete degradation of xenobiotic surfactants by consortia of aerobic microorganisms. Biodegradation 7:151–164.
Van Loosdrecht MC, Lyklema J, Norde W, Schraa G, Zehnder AJB (1987) The role of bacterial cell wall hydrophobicity in adhesion. Appl Environ Microbiol 53:1893–1897.
Van Toor RF, Hayes AL, Cooke BK, Holloway PJ (1994) Relationships between the herbicidal activity and foliar uptake of surfactant-containing solutions of glyphosate applied to foliage of oats and field beans. Crop Prot 13:260–270.
Varughese P, Broge A (1991) Base hydrolysis of ethylbenzoate in an oil-in-water microemulsions. J Indian Chem Soc 68:323–326.
Vera S, Rodenas E (1986) Inhibition effect of cationic micelles on the basic hydrolysis of aromatic esters. Tetrahedron 42:143–149.
Verma G, Kumar A, Tomar A, Singh KK (2004) Studies on persistence of temephos in sandy loam soil under field and laboratory conditions. J Environ Sci Eng 46:108–115.
Villedieu JC, de Savignac A, Calmon JP (1995) Kinetics and mechanisms of hydrolysis of dicarboximide fungicides in micellar media. J Agric Food Chem 43:1948–1953.
Volkering F, Breure AM, van Andel JG, Rulkens WH (1995) Influence of nonionic surfactants on bioavailability and biodegradation of polycyclic aromatic hydrocarbons. Appl Environ Microbiol 61:1699–1705.
Walters GW, Aitken MD (2001) Surfactant-enhanced solubilization and anaerobic biodegradation of 1,1,1-trichloro-2,2-bis(*p*-chlorophenyl)ethane (DDT) in contaminated soil. Water Environ Res 73:15–23.
Wamhoff H, Schneider V (1997) Photodegradation of imidacloprid. J Agric Food Chem 47:1730–1734.

Wan LSC, Lee PFS (1974) CMC of polysorbates. J Pharm Sci 63:136–137.
Wang CJ, Liu ZQ (2007) Foliar uptake of pesticide: present status and future challenge. Pestic Biochem Physiol 87:1–8.
Wauchope RD, Young JR, Chalfant RB, Marti LR, Sumner HR (1991) Deposition, mobility and persistence of sprinkler-irrigation-applied chlorpyrifos on corn foliage and in soil. Pestic Sci 32:235–243.
Werkheiser WO, Anderson SJ (1996) Effect of soil properties and surfactant on primisulfuron sorption. J Environ Qual 25:809–814.
White JC, Peters R, Kelsey JW (2007) Surfactants differentially impact *p,p*'-DDE accumulation by plant and earthworm species. Environ Sci Technol 41:2922–2929.
Willis GH, McDowell LL (1987) Pesticide persistence on foliage. Rev Environ Contam Toxicol 100:23–73.
Wolfe T, von Bünau G (1986) Photochemically induced viscosity changes in micellar cetyltrimethyl-ammonium bromide solution containing 4-hydroxystilbene. J Photochem 35:239–244.
Wolfe MF, Schwartz GJB, Singaram S, Mielbrecht EE, Tjeerdema RS, Sowby ML (1998a) Effects of salinity and temperature on the bioavailability of dispersed petroleum hydrocarbons to the golden-brown algae, *Isochrysis galbana*. Arch Environ Contam Toxicol 35:268–273.
Wolfe MF, Schlosser JA, Schwartz GJB, Singaram S, Mielbrecht EE, Tjeerdema RS, Sowby ML (1998b) Influence of dispersants on the bioavailability and trophic transfer of petroleum hydrocarbons to primary levels of a marine food chain. Aquat Toxicol 42:211–227.
Wolfe MF, Schwartz GJB, Singaram S, Mielbrecht EE, Tjeerdema RS, Sowby ML (1999) Influence of dispersants on the bioavailability and trophic transfer of phenanthrene to algae and rotifers. Aquat Toxicol 48:13–24.
Wolfe MF, Schwartz GJB, Singaram S, Mielbrecht EE, Tjeerdema RS, Sowby ML (2001) Influence of dispersants on the bioavailability and trophic transfer of petroleum hydrocarbons to larval topsmelt (*Arherinops affinis*). Aquat Toxicol 52:49–60.
Woodham DW, Hatchett JC, Bond CA (1974) Comparison of dimethoate and dimethoxon residues in citrus leaves and grapefruit following foliar treatment with dimethoate wettable powder with and without surfactant. J Agric Food Chem 22:239–242.
Xiaozhen F, Bo L, Aijun G (2005) Dynamics of solar light photodegradation behavior of atrazine on soil surface. J Hazard Mater B117:75–79.
Xiarchos I, Doulia D (2006) Effect of nonionic surfactants on the solubilization of alachlor. J Hazard Mater B136:882–888.
Xu J, Yuan X, Dai S (2006) Effect of surfactants on desorption of aldicarb from spiked soil. Chemosphere 62:1630–1635.
Yakata N, Sudo Y, Tadokoro H (2006) Influence of dispersants on bioconcentration factors of seven organic compounds with different lipophilicities and structures. Chemosphere 64:1885–1891.
Ye C (2003) Environmental behavior of the herbicide acetochlor in soil. Bull Environ Contam Toxicol 71:919–923.
Yeom IT, Ghosh MM, Cox CD (1996) Kinetic aspects of surfactant solubilization of soil-bound polycyclic aromatic hydrocarbons. Environ Sci Technol 30:1589–1595.
Ying GG (2006) Fate, behavior and effects of surfactants and their degradation products in the environment. Environ Int 32:417–431.
You G, Sayles GD, Kupferle MJ, Kim IS, Bishop PL (1996) Anaerobic DDT biodegradation: Enhancement by application of surfactants and low oxidation reduction potential. Chemosphere 32:2269–2284.
Zhang C, Zheng G, Nichols CM (2006) Micellar partitioning and its effects on Henry's law constants of chlorinated solvents in anionic and nonionic surfactant solutions. Environ Sci Technol 40:208–214.
Zhang Y, Miller RM (1994) Effect of a Pseudomonas rhamnolipid biosurfactant on cell hydrophobicity and biodegradation of octadecane. Appl Environ Microbiol 60:2101–2106.
Zhu L, Yang K, Lou B, Yuan B (2003) A multi-component statistic analysis for the influence of sediment/soil composition on the sorption of a nonionic surfactant (Triton X-100) onto natural sediments/soils. Water Res 37:4792–4800.

Index

MIX
Papier aus verantwortungsvollen Quellen
Paper from responsible sources
FSC® C105338

If you have any concerns about our products,
you can contact us on
ProductSafety@springernature.com

In case Publisher is established outside the EU,
the EU authorized representative is:
Springer Nature Customer Service Center GmbH
Europaplatz 3, 69115 Heidelberg, Germany

Printed by Libri Plureos GmbH
in Hamburg, Germany